SIMPLIFIED

■B■s of Advanced Mobile Communications

Jyrki T. J. Penttinen

Atlanta, GA, USA

Please follow the 5G blog at

http://www.5g-simplified.com

Contents, figures, and cover design: Jyrki T. J. Penttinen

Liability statement: The aim of this publication is to help readers to understand principles of the subject based on best efforts in summarizing the information. Regardless of all the careful work, including research, source studies, and educated statements, there always remains a possibility for errors and misinterpretations. Before any conclusions or actions such as a use, development, deployment or operation related to the described system, the reader is advised to ensure the correctness of this information from relevant standard bodies, authorities and other specialist sources. Thus, the author, or any other party related directly or indirectly to this book, will not assume any responsibility for possible health issues, loss of profit, or any other damage based on the contents of this book, and disclaim all warranties. The contents and statements presented in this book are solely of the author and do not necessarily represent the views of the current or past employers or any other entities. Furthermore, any organization, website, or product referred to in this work as a citation and / or potential source of further information does not mean that the publisher and author endorse the information or services.

©2019 Jyrki T. J. Penttinen

ISBN 978-1-0860-3260-4 (paperback)

Library of Congress Control Number: 2019911995

Imprint: Independently published

1st Edition 2019

Atlanta, GA, USA

Table of Contents

The Author

 Dr. Jyrki T. J. Penttinen, the author of **5G Simplified**, has worked in mobile communication industry since 1994, for Telecom Finland (Sonera and TeliaSonera Finland), Xfera Spain (Yoigo), Nokia and Nokia Siemens Networks in Mexico, Spain and in the United States, and G+D Mobile Security Americas. He is experienced in operational and research activities of system and architectural design, investigation, standardization, training, and technical management.

Since 2018, he has worked for GSMA North America as a Technology Manager assisting operator members with the adoption, design, development, and deployment of GSMA specifications.

Dr. Penttinen obtained his MSc (E.E.), LicSc (Tech) and DSc (Tech) degrees from the Helsinki University of Technology (currently known as Aalto University, School of Science and Technology) in 1994, 1999 and 2011, respectively. He also has been active lecturer and author of technical articles and telecommunications books.

Other Books of the Author

- ✓ **5G Explained:** Security and Deployment of Advanced Mobile Communications, Wiley, 2019

- ✓ **Wireless Communications Security:** Solutions for the Internet of Things, Wiley, 2016

- ✓ **The LTE-Advanced Deployment Handbook:** The Planning Guidelines for the Fourth Generation Networks, Wiley, 2016

- ✓ **The Telecommunications Handbook:** Engineering Guidelines for Fixed, Mobile and Satellite Systems, Wiley, 2015

- ✓ **The LTE / SAE Deployment Handbook**, Wiley 2011 (Mandarin edition in 2013)

- ✓ **The DVB-H Handbook:** The Functioning and Planning of Mobile TV (Penttinen, Jolma, Aaltonen, Väre), Wiley, 2009

Forewords

5G is without a doubt one of the hottest topics in current mobile communication industry. Thus, this is an excellent moment to study the most advanced cellular system to date to understand its possibilities. For the ones planning to work on the topic, 5G will provide interesting opportunities for years to come.

Realizing that 5G is such a fascinating base for many new, unheard applications, I decided to research the topic more. I published **5G Explained** with Wiley in 2019 and started to give related presentations to technical and non-technical audiences. I found that many were interested in somewhat lighter material to study the basics of 5G without spending too much time in such comprehensive technical foundations and terminology.

So, a publication which would present overall technical aspects of 5G in an easy-to-understand way started to make sense, and as a result, I wrote **5G Simplified** in the same year.

As the title indicates, the idea of this book is to demystify 5G technology in a common-sense yet concrete way, and to serve as an introduction for students and other interested parties with or without prior knowledge on mobile communications.

The focus of this edition is on some of the most important aspects, functionalities, and principles of the initial phase of 5G based on the 3GPP specifications. It also functions as an introduction to the topics presented in the *5G Explained* book, for the ones interested in more details.

I wanted to make this publication feasible for longer term purposes, too, as a practical and easy-to-use reference guide in eBook and printed formats. This book thus contains short "technical articles" of the most important 5G items in alphabetical order. Some of the chapters present technical principles such as Core Network, which summarizes the new 5G network functions, while other chapters are more generic such as Business Models. Depending on your current areas of interest, the modularity of this book should help familiarize you with the most relevant topics in any order you prefer.

Thanks to this modularity, I believe it is straightforward to capture the principles of the topics of interest without need to read through the whole book. At the end of each chapter, I have indicated the related key specifications of relevant standardization bodies for easing further studies. The references also indicate *5G Explained* sections where you can find more details.

I am happy for the feedback which motivated me to write this book. I would appreciate your further comments, questions and suggestions so please do not hesitate to contact me via the web site of this publication *http://www.5g-simplified.com* which also collects overall 5G news.

Last but not least, I want to thank my wife Celia, as well as Katriina, Pertti, Stephanie, Carolyne, Miguel, and all my close family and friends for their support. My very special thanks go to Campbell for the review.

I wish you a happy journey through the fascinating world of 5G, and I hope you enjoy this book as much as I enjoyed writing it!

Jyrki T. J. Penttinen

Atlanta, GA, USA

Abbreviations

3GPP	3rd Generation Partnership Project
5G	Fifth Generation mobile communication system
5GAA	5G Automotive Association
5GC	5G Core network
5GS	5G System
AF	Application Function
AI	Artificial Intelligence
AKA	Authentication and Key Agreement
AMF	Access and Mobility Management Function
AN	Access Network
API	Application Programming Interface
AR	Augmented Reality
ARPF	Authentication Credential Repository and Processing Function
AS	Application Stratum
ASN	Abstract Syntax Notation number one
AUSF	Authentication Server Function
BGA	Ball Grid Array

CAPEX	Capital Expenditure
CDMA	Code Division Multiple Access
C-IoT	Cellular IoT
CK	Ciphering Key
CN	Core Network
COTS	Commercial Off-the-Shelf
CPE	Customer Premises Equipment
CPU	Central Processing Unit
CUPS	Control and User Plane Separation
C-V2X	Cellular Vehicle to Everything
DANOS	Disaggregated Network Operating System
DFN	Dual Flat No Lead package
DFT	Discrete Fourier Transform
DN	Data Network
DNS	Domain Name System
EC	Extended Coverage (GSM IoT)
EIR	Equipment Identity Register
eLTE	evolved LTE
eMBB	evolved Mobile Broadband
EMC	Electromagnetic Compatibility
EMF	Electro Magnetic Field
eNB	evolved NodeB (LTE "base station")

EN-DC	E-UTRAN New Radio – Dual Connectivity
EPC	Evolved Packet Core (4G)
EPS	Evolved Packet System (4G)
eSIM	Electronic SIM
E-SMLC	Evolved Serving Mobile Location Centre
eSSP	Embedded SSP
ETSI	European Telecommunications Standardisation Institute
eUICC	Embedded UICC
FB	Fallback
FCC	Federal Communications Commission
FF	Form Factor
FG	Focus Group (ITU)
FR	Frequency Range
FWA	Fixed Wireless Access
GMLC	Gateway Mobile Location Center
gNB	Next generation Node B (5G "base station")
GNSS	Global Navigation Satellite System
GSM	Global System for Mobile Communications (2G)
GSMA	GSM Association
GST	Generic Network Slice Template
HO	Handover

HPLMN	Home PLMN
HS	Home Stratum
hSEPP	Home Security Edge Protection Proxy
HSPA	High Speed Packet Access
HW	Hardware
ICP	Internet Content Provider
ICT	Information and Computer Technologies
IK	Integrity Key
IMEI	International Mobile Equipment Identity
IMS	IP Multimedia Subsystem
IMT-2020	International Mobile Telecommunications 2020
IoT	Internet of Things
ISP	Internet Service Provider
iSSP	Integrated SSP
ITS	Intelligent Traffic System
ITU	International Telecommunications Union
ITU-R	Radio section of the ITU
iUICC	Integrated UICC
JSON	JavaScript Object Notation
JTAPI	Java Telephony Application Programming Interface
K	Permanent subscription key

KDF	Key Derivation Function
LBO	Local Breakout
LBS	Location Based Service
LMF	Location Management Function
LOS	Line of Sight
LPWAN	Low Power Wide Area Network
LTE	Long Term Evolution (4G)
LTE-M	LTE Machine-Type Communications
MAC	Medium Access Control
MCC	Mobile Country Code
MFF2	Machine-to-Machine Form Factor 2
MIMO	Multiple In, Multiple Out
ML	Machine Learning
mMTC	Massive Machine Type Communications
MNC	Mobile Network Code
MNO	Mobile Network Operator
MSIN	Mobile Subscription Identification Number
MTC	Machine Type Communications
N3IWF	Non-3GPP Interworking functions
NAS	Non-Access Stratum (dialogue between UE-core)
NB-IoT	Narrow Band Internet of Things
NEF	Network Exposure Function

NEST	Network Slice Type
NF	Network Function
NFV	Network Functions Virtualization
NG	Next Generation (5G interface)
NGC	Next Generation Core (5G), *See* 5GC
NG-RAN	Next Generation RAN (5G)
NR	New Radio (5G)
NRF	NF Repository Function
NSA	Non-Standalone
NSSF	Network Slice Selection Function
NWDAF	Network Data Analytics Function
NWM	Network Management
OCP	Open Compute Project
OFDM	Orthogonal Frequency Division Multiplex
OOB	Out of Band
OPEX	Operational Expenditure
OTA	Overt the Air
OTDOA	Observed Time Difference of Arrival
P2MP	Point-to-Multipoint
PAPR	Peak-to-Average Power Ratio
PCF	Policy Control Function
PDCP	Packet Data Convergence Protocol

PDN	Packet Data Network
PDU	Packet Data Unit
P-GW	Packet Data Network Gateway
PLMN	Public Land Mobile Network
PON	Passive Optical Network
PRS	Positioning Reference Signal
PS	Packet Switched
QAM	Quadrature Amplitude Modulation
QoE	Quality of Experience
QoS	Quality of Service
QPSK	Quadrature Phase Shift Keying
RAN	Radio Access Network
RAT	Radio Access Technology
RCS	Rich Communications Services
REST	Representational State Transfer
RF	Radio Frequency
RLC	Radio Link Control
RoI	Return of Investment
RPMA	Random Phase Multiple Access
RRC	Radio Resource Control
RRM	Radio Resource Management
RRU	Radio Remote Unit

RSU	Road Side Unit
SA	Standalone
SAP	Service Access Point
SAR	Specific Absorption Rate
SAS	Security Accreditation Scheme
SC	Single Carrier
SCEF	Service Capability Exposure Function
SCMF	Security Context Management Function
SDN	Software Defined Networking
SDO	Standard Development Organization
SDSF	Structured Data Storage Function
SEAF	Security Anchor Function
SEPP	Security Edge Protection Proxy
S-GW	Serving Gateway
SIDF	Subscription Identifier De-Concealing Function
SIM	Subscriber Identity Module
SLA	Service Level Agreement/Assurance
SLP	SUPL Location Platform
SMF	Session Management Function
SMSF	Short Message Service Function
SoC	System on Chip
SPCF	Security Policy Control Function

SS	Serving Stratum
SSP	Smart Secure Platform
SUCI	Subscription Concealed Identifier
SUPI	Subscription Permanent Identifier
SUPL	Secure User Plane
SW	Software
TA	Tracking Area
TDMA	Time Division Multiple Access
TP	Transmission Point
TR	Technical Report
TS	Technical Specification
TS	Transport Stratum
TSON	Time Shared Optical Network
UDM	Unified Data Management
UDR	Unified Data Repository
UDSF	Unstructured Data Storage Function
UE	User Equipment
UICC	Universal Integrated Circuit Card
UMTS	Universal Mobile Telecommunications System (3G)
UPF	User Plane Function
URLLC	Ultra-Reliable Low Latency Communications

USIM	Universal Subscriber Identity Module
V2I	Vehicle to Infrastructure
V2N	Vehicle to Network
V2P	Vehicle to Pedestrian
V2V	Vehicle to Vehicle
VM	Virtual Machine
VoLTE	Voice over LTE (4G)
VoNR	Voice over New Radio (5G)
VPLMN	Visited PLMN
VR	Virtual Reality
vSEPP	Visited network's Security Edge Protection Proxy
WDM	Wavelength Division Multiplexing
Wi-Fi	WLAN, based on the Institute of Electrical and Electronics Engineers' (IEEE) 802.11 standards
WLAN	Wireless Local Area Network (Wi-Fi)
WRC	World Radiocommunication Conference

ABCs – Introduction

5G refers to the 5th generation of mobile communication systems. It represents mobile telecommunication standards beyond 4G LTE, and it will comply with the strict IMT-2020 (International Mobile Telecommunications 2020) requirements of ITU-R (Radio section of the International Telecommunications Union).

5G provides much faster data speeds with very low latency compared to the legacy systems. 5G also supports a higher number of devices communicating simultaneously.

5G is capable of handling much more demanding mobile services than was ever possible before, including tactile Internet and virtual reality applications which will provide completely new and highly attractive user experiences.

As LTE and its evolution, LTE-Advanced, have been a success story serving a growing base of customers, one might thus ask why we need yet another generation? Simply, the answer is the same as before: the older generation's performance is reaching a critical limit and it makes more sense to define a new system from scratch instead of trying to enhance legacy platforms.

Capacity has oftentimes been one of the biggest limitations as the customer base and demand for data consumption continue to grow. Not only is this an issue for us as consumers, but the increasing popularity of Internet of Things (IoT) devices in impacting on the system design as networks need to support a massive number of intelligent sensors and other devices relying on Machine Type Communications (MTC).

The standardization community foresaw this trend and defined a new generation of specifications. They support up-to-date performance figures which would be challenging to comply with by adding on and developing the previous generations further. As a result, 3GPP (3rd Generation Partnership Project) released the first phase 5G Technical Specifications (TS) in the beginning of 2019 with a completely renewed system architecture.

The new era of connected society is quickly becoming a reality as operators start to deploy the initial 5G networks. GSMA is estimating that 5G will account for 15% of the global mobile industry by 2025 as stated in the Mobile Economy 2019 study. (GSMA, 2019) In other words, we will reach 1.4 billion 5G connections within the next 5 years.

Although industry is eager to offer 5G services already in expedited schedules, it will be some time until we can enjoy the performance of the full version of 5G. Although the current 3GPP Release 15 sets the scene, it only works as an introduction to 5G as ITU envisions it.

Later in 2019, 3GPP will produce Release 16 specifications and give industry the means to deploy the second phase 5G networks that will finally comply with the ITU's rather demanding IMT-2020 requirements for the global 5G. Not only will it offer superior data speeds outperforming all the previous generations, but the full version of 5G also introduces advanced solutions such as network slicing, virtualized network functions, support for massive amount of IoT devices communicating simultaneously, ultra-reliable and low-latency communications, edge computing, and optimized service-based architecture model. The security architecture is renewed, too, to better cope with modern cyber-attacks.

So, the current, early stage of 5G offers merely a taste of *evolved mobile broadband* service. While waiting for the advanced features of the next phase 5G to become a reality, consumers can start to feel the benefits of evolving 5G services thanks to the deployment of 3GPP's intermediate options which provide practical Non-Standalone scenarios combining 4G and 5G elements, thus easing the roll-out of the new networks.

Figure 1 summarizes some of the key 5G functionalities. There are also other solutions increasing 5G performance such as enhanced radio interface modulation and intelligent multi-array antenna solutions. As soon as the new 5G radio frequency (RF) bands are deployed especially on the mm-Wave spectrum, the networks can offer more generous capacity to support wideband radio transmission, which in turn provides users with clearly increased data speeds.

The next World Radiocommunication Conference of ITU later in 2019 will play an important role to provide concrete frequency band allocation plans at a global level. (ITU, 2019) New frequencies will pave the way for increased capacity needed as people start using advanced 5G applications. At the same time, we are expecting new services that will change how we communicate.

5G will include room for *network slicing* which refers to a set of optimized "networks within network". Depending on the specific needs of the verticals, which are the practical representatives of varying communication profiles such as drones, law enforcement, automotive, self-driving vehicles, critical infrastructure and smart cities, each one may obtain enough of a suitable network slice to fulfil their specific communications needs.

As an example, a network slice can provide its users with a very high data speed while some other slice can offer low data speeds but high reliability, etc. The previous systems were not able to distinguish between the offered services in this way.

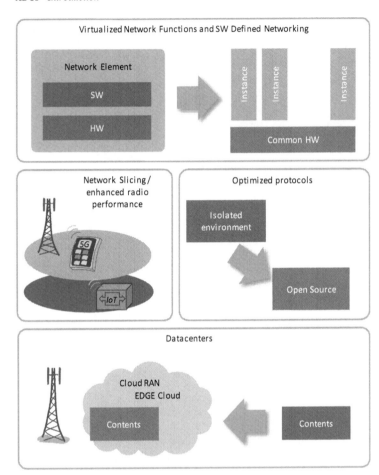

Figure 1 Key functionalities of 5G.

5G is also based increasingly on Open Source concept which generates business models and provides opportunities for many new stakeholders to join the developer community.

In practice, 5G will evolve gradually. It will provide enhanced performance as the network deployments continue and the coverage extends. Meanwhile, the previous generations can serve mobile customers for years to come as 5G experiences inevitable outages, especially in the beginning.

5G will offer the most significant benefits in dense urban areas where small cells and mm-Wave RF bands provide the highest data speeds. Also, the closer to the users the edge computing is located, the lower the latency. This particularly benefits applications that are delay-sensitive, such as autonomous cars.

The downside of mm-Wave cells is their highly limited coverage area due to the increased attenuation on higher frequencies. In fact, 5G small cell may be in range for only a few hundred feet without obstacles in the communication line. The construction of such a dense radio network also requires connectivity to the high-capacity core infrastructure. Thus, this deployment strategy is only feasible in limited areas such as a city center where the antennas could be installed broadly on structures such as light poles. In sub-urban and rural areas, 5G can rely on lower frequency bands which serve much larger coverage areas. While the achieved performance is somewhat lower, it will be much better than any previous generation in the same area.

5G will facilitate new commercial models and business opportunities for growing number of greenfield stakeholders. One example of new 5G-based businesses is related to data centers which serve as a useful platform to process new 5G network functions. Not all operators may be interested in investing in this type of infrastructure, at least in the beginning. This provides new business opportunities for cloud service providers, too.

Despite the highly advanced and complex technology, there are expectations for decently priced consumer devices. According to the latest indications of the industry, we may see 5G smart devices in 300 US dollar category rather soon. There will be market for many types of devices, though, from simple 5G-connected sub-10 US dollar sensors up to complex, high-end virtual reality headsets which could be capable of processing and transferring 360-degree, 3D audio-video contents.

The following chapters present some of the most important aspects of 5G in alphabetical order to ease the studies of 5G. The contents are generic yet technically concrete, so there is no need for previous knowledge on mobile communications. Furthermore, this book is designed to be a "pocket dictionary" of 5G, so the following chapters are completely independent from each other and can be read through in any order to understand the most interesting items.

Lastly, for those interested in more details by reading standards, each chapter presents a couple of the most important references as shown below. The eBook format of this publication has embedded links to the referenced 3GPP specifications. The readers of the printed book can find the specifications at the web page below replacing the *xxxxx* with the respective specification number without a dot; as an example, the 3GPP TS 33.501 would be written as "33501".

https://www.3gpp.org/DynaReport/xxxxx.htm

This link leads you to the specification's summary page which shows the relevant generations it relates to. By selecting a link on that page, "*Click to see all versions of this specification*", you will get the complete list of releases and versions of the specification of interest. To examine the very latest version, you can download the one with the highest version number.

Further reading	
▶3GPP	*TR 22.891* (new services and markets technology enablers)
▶ITU	*IMT-2020* (International Mobile Telecommunications)
▶5G Explained	Section 6.1 (core network overview)

Architecture

The service-based architecture of 5G refers to the capability of the network to present its elements as network functions in virtualized environment. It enhances the performance of the network and provides means for new features.

The new 5G architecture model provides services between Network Functions (NF) using a common hardware. In the previous generations, dedicated stand-alone network elements were used for each network function such as the LTE Serving Gateway (S-GW).

5G architecture model benefits from the high performance of *Network Functions Virtualization* (NFV) and *Software-Defined Networking* (SDN). As a result, 5G network is faster, more reliable, and easier to manage compared to the older networks. Network Repository Functions (NRF) have a special role in 5G architecture as they allow the network functions to discover the services of the other network functions in a secure manner.

Furthermore, 5G is based on *Network Slicing*. It is a set of features that form a complete, virtual cellular network, a Public Land Mobile Network (PLMN). It connects smart phones and other 5G devices referred to as User Equipment (UE). While traditional architectures define only one uniform, physical cellular network, 5G network slicing provides means for a single

MNO (Mobile Network Operator) to form a set of parallel PLMNs within the same physical area. In other words, slices can be considered as "networks within network".

MNO can set up and optimize each slice individually to achieve the best performance for different usage scenarios "on-the-fly". Slices can be created and terminated when needed, temporarily or for longer time periods. As an example, operator can set up a specific slice to provide fast data for its users while some other slice may serve a huge amount of low bit-rate sensors within the very same service area. Adjusting adequately the functional and performance parameters, different users benefit from the selection of the most suitable slices. This enhances the user experience, and MNO has a better possibility to optimize the offered network capacity.

Furthermore, 5G has a novel Quality of Service (QoS) concept which can better cope with differentiating the data flows based on varying priority levels of services.

Yet another benefit of 5G is the support of a variety of access systems. In addition to the new 5G radio network itself (Next Generation Radio Access Network, NG-RAN), 5G core network (5GC) can serve also generic access networks (AN) such as Wi-Fi hotspots. In the beginning, the 5G core network can be connected to both 3GPP NG-RAN and 3GPP-defined untrusted WLAN (Wi-Fi) whereas the next Release 16 will include more access options.

The 3GPP Release 15 defines the initial 5G system architecture and functionalities. Some of the most important technical specifications (TS) are 3GPP TS 23.501, TS 23.502 and TS 23.503 which also describe the evolved Mobile Broadband (eMBB) data service, subscriber authentication and authorization, application support, edge computing, IP Multimedia Subsystem

(IMS), and interworking with 4G and possible other access systems.

The 3GPP Release 16 provides many new 5G features. As an example, 5G will integrate massive Machine Type Communications (mMTC) into 5G as a continuum from the Narrow-Band IoT (NB-IoT) and LTE-M (LTE Machine-to-Machine) services of the previous LTE releases. The Release 16 also enhances the Ultra-Reliable Low Latency Communication (URLLC) mode that reduces latency.

With all the Release 16 enhancements, 5G will be able to comply with the strict requirements of the ITU IMT-2020 (International Telecommunications Union, International Mobile Telecommunications). As an example, the Release 16 provides as low latency values as 1 ms thanks to the URLLC mode and edge cloud technology. They facilitate the moving of contents and processing it closer to the users, minimizing the delays.

The Release 16 includes around two-dozen additional items such as Multimedia Priority Service, Vehicle-to-Everything (V2X) application-layer services for car-to-car, ship-to-ship and railway communications, 5G satellite access, LAN support, convergence of wireless and wireline, enhanced terminal positioning, network automation, and evolved radio techniques, security, codecs and streaming services. (3GPP, 2018) All these are beneficial for a connected society using 5G as a platform.

The work does not end here. The 3GPP working groups are already assessing new features for the forthcoming Release 17 which keeps enhancing 5G. 3GPP will have approved these work items by December 2019. (3GPP, 2018)

The 3GPP Release 15 refers to as the *first phase* 5G, and the Release 16 represents the *second phase*. The architectural options of 5G include Standalone (SA) and Non-Standalone

(NSA) deployment scenarios which allow phased deployment. It means that 5G operators may select their initial network strategy relying on supportive 4G infrastructure, and change the architecture model to fully 5G-based as specifications mature and respective devices will be available. Please see Deployment Chapter for more details on these NSA and SA options.

The 3GPP TS 38.401 defines a *New Radio* (NR) architecture and the respective interfaces. It describes a logical separation of signaling and user data transport. The separation can be deployed within a 5G network or between the new 5G and legacy 4G networks. In this way, the signaling load and user data capacity can be optimized. For more details on the 5G radio, please refer to Radio Network and Frequencies Chapters.

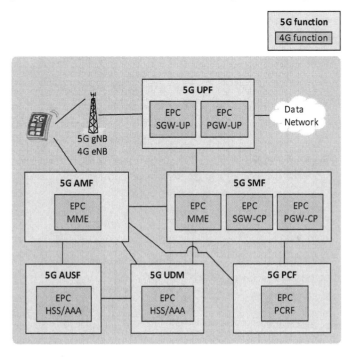

Figure 2 The mapping of the most important 4G and 5G elements.

Each 5G network function performs tasks as described in the 3GPP TS 23.501. Figure 2 depicts the mapping of the 4G and 5G key functions, and Table 1 compares their roles. In Figure 2, the EPC refers to Evolved Packet Core network of 4G, eNB to 4G base station (evolved NodeB), and gNB to 5G base station (next generation NodeB). The service-based architecture provides gradual 5G network deployments, and MNOs can always take advantage of the latest advancements of the virtualization concept.

Table 1 The key elements of 5G, and mapping with the 4G LTE system.

5G NF	Description	Mapping with 4G
5G-EIR	Equipment Identity Register	Evolution of LTE Equipment Identity Register (EIR)
AF	Application Function	LTE Application Server (AS) and GSM Service Control Function (gsmSCF)
AMF	Access and Mobility Management Function	Replaces the LTE Mobility Management Entity (MME)
AUSF	Authentication Server Function	Replaces the LTE MME / AAA (Authentication, Authorization and Accounting)
NEF	Network Exposure Function	Evolution of Service Capability Exposure Function (SCEF) and API layer
NRF	NF Repository Function	Part of the evolution of Domain Name System (DNS)
NSSF	Network Slice Selection Function	New function for 5G-specific network slicing concept (not in 4G)
PCF	Policy Control Function	Evolution of the LTE Policy and Charging Enforcement Function (PCRF)
SEPP	Security Edge Protection Proxy	New element for securely interconnecting 5G networks (not in 4G)
SMF	Session Management Function	Replaces, with the 5G UDF, the LTE Serving (S-GW) and PDN Gateway (P-GW)
UDM	Unified Data Management	Evolution of Home Subscriber Server (HSS) and Unified Data Repository (UDR)
UDR	Unified Data Repository	Evolution of the LTE Structured Data Storage (SDS)
UDSF	Unstructured Data Storage Function	The function comparable with LTE Structured Data Storage Function (SDSF)
UPF	User Plane Function	Replaces, with the SMF, the LTE S-GW and P-GW

Some functional 5G elements differ considerably from their counterparties in LTE increasing performance and adding functionalities. Perhaps the most significant changes are related to the original S-GW (Serving Gateway) and P-GW (PDN Gateway, or Packet Data Network Gateway) of the LTE core network, referred to as S/P-GW in their collocated form. Those elements are the "motor" of the 4G core network as they deliver the data between LTE devices and other destinations.

In 5G, the tasks of those elements are divided into a separate user plane (UP) for the data connectivity via UPF (User Plane Function), and into a control plane (CP) for the signaling via SMF (Session Management Function).

Also, rather important change is made to Authentication, Authorization and Accounting (AAA) and Home Subscription Server (HSS) elements of 4G; they are located to the Authentication Server Function (AUSF) and Unified Data Management (UDM) in the 5G architecture model.

Please see further details of the 5G network functions in Core Chapter.

Further reading	
▶3GPP	*TS 23.501* (system architecture); *TS 23.502* (procedures); *TS 23.503* (policy and charging)
▶5G Explained	Chapter 4 (architecture)

Building Blocks of 5G: eMBB/URLLC/mMTC

5G Core (5GC) and New Radio (NR) form the 5G System (5GS). It enhances capacity and performance, and complies with the most demanding requirements for evolved mobile broadband (eMBB), massive machine type communications (mMTC), and ultra-reliable low latency communications (URLLC).

The eMBB, mMTC, and URLLC are the elemental building blocks, or "dimensions", of 5G as indicated in the ITU IMT-2020 which sets the reference for the new era. (ITU-T, 2017) Figure 3 summarizes their respective capabilities.

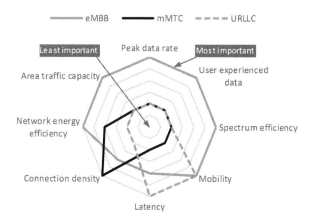

Figure 3 5G dimensions and their capabilities as per ITU IMT-2020. As an example, the connection density is the most important capability for mMTC whereas the lowest possible latency is of utmost importance for URLLC.

Figure 4 depicts these dimensions. It also presents additional remarks of practical, expected 5G service usage environments as interpreted from the Technical Report of 3GPP, TR 22.891 (Release 14). It details 74 feasibility studies on new 5G services and technology enablers of the market.

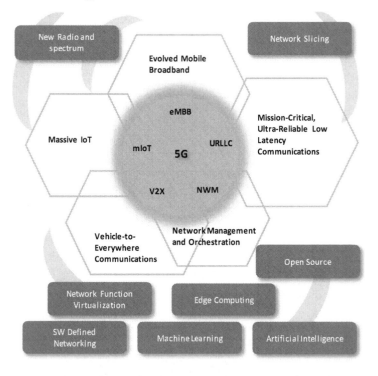

Figure 4 The building blocks of 5G. The eMBB, URLLC, and mMTC are the main dimensions and usage types of 5G as defined by the ITU IMT-2020.

5G is able to support a variety of *use cases* independently from each other. Each of these use cases require different performance values in terms of area traffic capacity, peak data speed, user's experienced data speed, spectrum efficiency, mobility, latency, connection density, and energy efficiency. These characteristics can be referred to as attributes, and their values dictate their capabilities.

The offering of these attributes with an adequate quality of service (QoS) is possible thanks to the new 5G technologies such as virtualized 5G network functions, software-defined networking, edge computing, and network slicing. Please see more details of these topics in Core and Network Slicing Chapters.

Nevertheless, as 5G is still in its initial deployment phase and has rather fragmented coverage in the beginning, some of the use cases might not be quite yet ready for the verticals requiring high reliability and performance within large areas.

URLLC: This dimension provides the lowest latency values and serves thus well for fulfilling the needs of critical communications. Some potential use cases of this category include drone flight control, localized factory automation, and wide-area smart city functions. Although the base for the URLLC framework was identified already in the 3GPP Release 15 NR, the Release 16 adds and enhances further the URLLC features.

mMTC: This dimension is designed to support a huge amount of simultaneously communicating Internet of Things (IoT) devices and applications. The LTE Releases 13 and 14 define already low-power wide area (LPWA) Cellular IoT (C-IoT) modes which are NB-IoT and LTE-M. It can be assumed that these already existing modes comply largely also with the 5G mMTC requirements whereas the Release 16 will enhance their performance to ensure that full compliancy is achieved with the IMT-2020 expectations. (GSMA, 2018)

Table 2 Description of the 5G dimensions.

Dimension	Description
URLLC	For the highest reliability and lowest latency applications
mMTC	For the densest sensor and other IoT environments
eMBB	For the highest data rate applications
Management	Evolved, machine learning -based self-optimized network
V2X	5G can provide base for vehicle communications

eMBB: This dimension provides very high data speeds clearly exceeding the 4G speeds, and is included in the 5G system as of the Release 15. The respective first phase deployments focus on this category until URLLC and mMTC will be fully available later in the Release 16 timeframe. Even the Release 16 deployments enhance further the 5G performance, this intermediate phase already paves the way for much more demanding mobile broadband use cases such as AR/VR applications.

As the network functions become more complex and dynamic, it can be assumed that the **Network Management and Orchestration** will rely increasingly on evolved Self-Optimizing Network (SON) concept, artificial intelligence (AI), and machine learning (ML). 3GPP has designed and enhanced SON already for previous generations, and the same principles apply still in 5G era although the network slicing with its high dynamics requires even more advanced versions of SON.

V2X, Vehicle-to-Everywhere, is a good example of an environment where some of the most critical 5G performance requirements are needed, especially if self-driving vehicles are involved. It also is an example of an automotive vertical that would benefit from the maximum performance that is a combination of all the dimensions, i.e., URLLC, mMTC and eMBB.

In addition to the 5G-based car communications, 3GPP is extending Vehicle-to-Vehicle (V2V) to cover also special environments such as maritime communications. As GSM and its GSM-R (Railway) communications mode may gradually turn out to be obsolete along with the probable sunset of many 2G networks, 5G would be, in fact, quite logical base for the future railway communications and control platforms.

Further reading	
▶3GPP	*TR 22.891* (new services and markets technology enablers)
▶ITU	*ITU-R Rec. M.2038* (IMT for 2020 and beyond)
▶5G Explained	Section 6.1 (core network overview)

Business Models

5G generates plenty of new businesses. As it is based on novelty solutions such as cloud infrastructure and virtualized network functions, it also opens doors for many new stakeholders. One example of these is the data center service providers.

Some of the primary tasks of mobile network operators (MNO) are to invest in the network infrastructure and to set up connectivity for services. On the other hand, the subscription fees generate the principal profit for the operators.

Before building up a 5G cellular network, MNO needs to obtain an operating license for one or more frequencies. This typically takes place via auctions that national frequency regulators, such as US-based FCC (Federal Communications Commission), organize. The terms of the license dictate the duration of the right to offer services on that band, and they detail also the allowed technologies the licensee may use.

Table 3 summarizes the results of one of the 5G auctions held in Germany on June 2019. In this case, 4 MNOs obtained their own licenses for mid-band frequencies, and the cost of each license was in range of 1-2 Billion euros.

The next major step of the MNO is to invest in and deploy the 5G infrastructure which may include radio, transport, and core networks. Instead of investing in the complete network, the

MNO may also consider leasing the needed capacity from transport and core service providers. In 5G, these networks are typically based on data centers.

The initial investment generates Capital Expenditure (CAPEX) while the costs of maintaining the infrastructure and services are referred to as Operating Expenditure (OPEX). The business model needs to be designed accordingly for the immediate and long-term investments and gains.

Table 3 Example of 5G frequency band auction. (Mobile World Live, 2019)

MNO	B euros	GHz	Capacity
Deutsche Telecom	2.2	2	4 blocks
		3.6	9 blocks
Vodafone	1.9	2.1	2 blocks of 15 MHz
			2 blocks of 5 MHz
		3.6	90 MHz
Telecom Deutschland	1.4	2.1	2 blocks
		3.6	7 blocks
1&1 Drillisch	1.1	2	2 blocks of 5 MHz
		3.6	5 blocks 10 MHz

A good deployment strategy of the 5G radio base stations (gNB, next generation Node B elements) is important. Taking into account immediate and future expenses, MNOs need to balance the techno-economic aspects of their radio network using three principal dimensions which are: the offered capacity, radio coverage, and quality of service (QoS).

Over-dimensioning the capacity, coverage, or QoS compared to the foreseen near-future outlook is typically unnecessarily expensive and may impact negatively on the business.

On the other hand, if the coverage or capacity and thus the quality are not adequate, it could easily increase customer churn as unhappy users may transport their subscriptions to other operators. Losing an existing customer is typically very expensive from operator's perspective; not only for the lost revenue of the subscription, but it may impact negatively even on the business image along with the easiness to present critical feedback in social media. So, the proper 5G radio network optimization is one of the operator's priority tasks.

MNOs typically divide radio network planning into *nominal* and *detailed* phases. Based on a high-level technical network modeling and radio link budget, the aim of the nominal phase is to get a first-hand estimate on the needed number of base stations and other investment items in the planned area taking into account the future outlook of the number of the customers. This phase is the foundation of the tecno-economic feasibility evaluation because the results may reveal if the business will be profitable, or if the commercial models need to be developed further to ensure the expected revenue.

This initial technical information translates into an estimate of the expected investment figures based on a cost breakdown of different base station types. The estimate may include a physical site shelter, equipment, antennas, and transmission lines in different geographical locations such as urban and rural areas. MNOs can obtain fairly accurate estimate on the total radio network deployment cost by multiplying these values with the overall site number in different geographical clusters.

Figure 5 presents a conceptual example of an outcome of a radio link budget (RLB) analysis for rural 5G cells. Among the other radio aspects, the RLB takes into account the effective radiating power of the base station antenna beams and the path loss by applying an appropriate prediction model for each area type. Based on the RLB, MNO can estimate the maximum feasible distance between the base station and the connected 5G device at the cell edge. This distance varies depending on the used service type (such as voice call or high-speed video) which must be taken into account in the plan.

The RLB results in the ideal distance between neighboring base stations and thus in the minimum number of the base stations needed in the area. It is important to include sufficient (but not too much) overlapping coverage in cell edges. In this case, an area of 300 km² can be covered by ten base stations so that the area coverage in the designed cluster is 90 %.

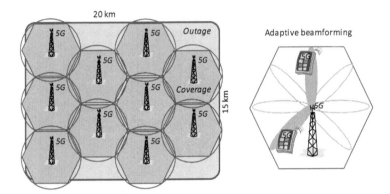

Figure 5 Example of the nominal radio network planning. In 5G, the coverage is formed typically by an array of adaptive beams instead of static cells.

If a single base station cost, including equipment, shelter and other expenses is, say, 100,000 US dollars, the CAPEX for this specific area is 1 Million USD. The possibly existing base station infrastructure of the MNO can be reused partly also for 5G which helps to reduce the CAPEX. Even if this exercise is rather simple and the cost of the realistic base stations may vary considerably, it is easy to guess that the deployment of a nationwide radio network could represent a major investment for any operator.

The yearly transmission, power consumption, maintenance and other OPEX items can be estimated in the same manner by multiplying a single base station's OPEX by the total number of the sites. It should be noted that the overall OPEX might need to include yearly frequency license fee, too.

The MNO needs to understand also the rest of the infrastructure and service costs such as core and transport, SIM card supply chain, and possibly subsidized mobile phones. In addition, the operational costs include plenty of other direct and indirect expenses such as human resources, operational tools, services and maintenance. It is also worth mentioning that network sharing may be a feasible business model for MNOs and Mobile Virtual Network Operators (MVNO), the latter renting capacity from the MNO's radio network and offering it to their customers.

The Return of Investment (RoI) depends largely on the gain generated from the operator's subscriptions. The MNO may also consider supportive revenue-generating businesses, taking advantage of the 5G infrastructure. One example is a 5G MNO combining cellular operations and tv broadcast business. The ultimate goal is to generate sufficient revenue stream and margin to support all the business areas of the company such as R&D and marketing while they help generating future business.

For the mobile equipment and device providers, the business model is related to the cost of research and development, manufacturing, selling, and distribution of the equipment such as network elements, smart phones, and IoT devices, or offering of managed services.

The optimization of the Bill of Material (BoM) plays a key role in the equipment business. An example of the related challenges is the important task to select the supported frequency bands of a new mobile device model. 3GPP specifications include many bands to select from, but the more the device supports, the more complex, large, heavy, and expensive it gets due to the number of additional components such as antennas, filters and switches.

There is a growing number of 5G frequency bands as indicated in Figure 13 of Frequencies Chapter. It is challenging to include all of them into a single device, so the manufacturers typically vary the models for regional and global markets. The aim is to select adequate set of useful bands. This is an example of the complexity of the device business which differentiates between regional, global, and specially tailored MNO-exclusive models.

The device manufacturers do not tend to detail the cost breakdown figures of the devices. Nevertheless, it could be assumed that the BoM of a high-end smart device may represent around one-third of its retail price according to, e.g., Ref. (Sherman, 2013). On the other hand, the monetization of Intellectual Property Rights of the essential functions may also play an important role in the business as indicated in Patents and IPR Chapter.

After the research, development, BoM, manufacturing, distribution, marketing, and other costs, the goal of the manufacturer is to achieve sustainable revenue. The return of investment has typically been the most profitable for the high-end smart device category, so this principle is probably valid also in 5G era.

The same business aspects are applicable to other manufacturers, too, such as SIM card, chipset and module vendors; their aim is to achieve feasible business for the products in different regions. For other stakeholders in the 5G ecosystem, such as service providers and data center operators outsourcing their capacity to MNOs, the business is related to the setting up of the services including potential service level agreements to set the performance expectations adequately.

Further reading	
▶3GPP	*TR 32.851* (OAM aspects of network sharing); *TR 38.913* (scenarios and requirements for next generation access); *TR 22.830* (Study on Business Role Models for Network Slicing)

Cloud RAN and Core

Cloud RAN (Radio Access Network), shortly C-RAN, is a centralized radio access network architecture model based on cloud computing for radio access networks. Equally, the core network can be based on cloud. Both support current mobile communication systems and future wireless standards.

Historically, each mobile communication radio base station has served users providing coverage areas individually and relying on their own, local capacity. The traditional radio networks are dimensioned based on a busy-hour blocking probability. The busy hour of the network refers typically to the most loaded 60 minutes of a complete 24-hour day.

The capacity of each base station is dimensioned based on this peak traffic so that the maximum blocking probability does not exceed the desired value. It is inevitable that small amount of the customers experience service degradation or congestion during the busy hour. The dimensioning of the average blocking probability is one of the many optimization tasks of the operators in order to balance the investments and adequate user experiences. The average load is typically much lower which results in unused capacity reserve during the off-peak hours.

In the traditional architecture model, the processor resources of base stations cannot be shared with others. The processing of the baseband radio signals happened within the same site as the power amplifier was located to. Additional challenge was the relatively long coaxial cables which were deployed from the transmitter to the antennas, resulting cable losses – which, in turn, impacted negatively on the radio link budget and costs.

Ever since, the radio base stations have been optimized by applying more sophisticated solutions. In the distributed base station model, the base stations rely on separation of the remote radio head (RRH) and the baseband unit (BBU) which are connected by fiber. These elements can be located rather far away from each other as the fiber optics has minimal signal loss compared to the coaxial cable. If the RRH is next to the antenna, the respective coaxial cable loss is insignificant. Also, as the remotely located BBU processing can be shared amongst more than a single base station, the overall capacity can be maximized. These solutions popular already in the 3G deployments.

C-RAN is the next step in this evolution. It can be based on up-to-date, or evolved CPRI (Common Public Radio Interface) standard which serves as a protocol in the respective digital baseband data streaming. The Coarse / Dense Wavelength Division Multiplexing (CWDM / DWDM) is used in the transmission on mm-Wave bands for the baseband signal over long distance wirelessly from the base station to the core network.

C-RAN is a cost-efficient, centralized base station deployment model relying on data center infrastructure. This model also provides 5G operators with a wide bandwidth access to the cloud BBU pool with a high reliability and low latency. The concept includes fluent dynamic resource sharing in multi-vendor, multi-technology environments. (Murphy, n.d.)

5G thus benefits from the C-RAN concept, and it can be deployed via fronthaul transmission interfaces as depicted in Figure 6. (Bougioukos, 2017) In addition, if operator so decides, 5G can still use the more traditional distributed and centralized RAN architecture models.

Both radio and core as well as the transport networks of 5G can be deployed on cloud, facilitated by the Software-Defined Networking (SDN) and Network Functions Virtualization (NFV). Cloud supports a variety of diversified services. In addition, cloud is the foundation for the network slicing which provides fast and efficient deployment of many new 5G services.

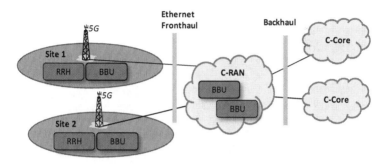

Figure 6 Cloud-based 5G Radio Access Network architecture model.

The C-RAN can be connected via 3GPP RAN or other, non-3GPP Access Network (AN) sites. Examples of the former are the 3GPP base stations (gNB of 5G, and base stations of also any earlier generations), and Wi-Fi access points represent the latter type. In addition to the access connectivity, the C-RAN architecture has mobile cloud engines (MCE) which coordinate services of the real time and non-real-time resources.

The C-RAN connects to the service-oriented Cloud Core (C-Core) network as depicted in Figure 7. The Cloud Core network has various tasks such as composable control function that manages mobility, services, policies, security, and user data. It also takes care of dynamic policy control of the supported services, and it manages the storing of data in the unified database.

Furthermore, the C-RAN has a control plane with a set of components for performing network functions, and programmable user plane fulfilling a variety of service requirements. The concept provides a possibility to orchestrate the network functions in order to select control and user plane functions.

As depicted in Figure 7, the SDN controller is in the transmission section between the C-RAN and Cloud Core. The SDN decouples the control and data planes, and centralizes the network control functions which optimizes the network management.

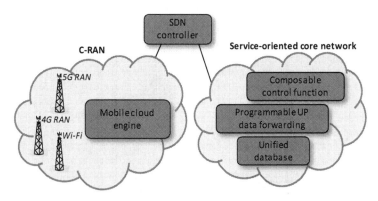

Figure 7 The principle of C-RAN and Cloud Core of 5G.

One of the key benefits of the SDN-enabled C-RAN deployment is the adaptive reconfiguration. In this mode, the SDN controller receives the status changes of the RAN in real time which enhances the performance of mobility management and load balancing of the BBUs when 5G users move within the service area and generate varying data transmission patterns.

Further reading	
▶ETSI	*White Paper 23. Cloud RAN and MEC: A Perfect Pairing*
▶ITU	*Technical Report*, GSTR-TN5G: Transport network support of IMT-2020/5G
▶5G Explained	Section 6.4.5 (Cloud RAN)

Core Network

5G changes the philosophy of the previous network models to bet-ter cope with today's communication environment. As a conse-quence, the old architectures are now replaced by service-based architecture model which virtualizes the network functions.

The virtualized 5G *Network Functions* (NF) replace the old model which has been relying on dedicated hardware elements. Figure 8 depicts the 5G Network Functions in the new service-based architecture (SBA) model for roaming.

In this scenario, the User Equipment (UE), such as smart phone or IoT device, roams visited network (VPLMN). The UE estab-lishes a connection to the Data Network (DN) while the Home Public Land Mobile Network (HPLMN), which is the user's home 5G network, enables its connectivity. The visited and home network communicate with each other for the user's sub-scription information (via UDM), subscriber authentication (via AUSF), and policies (via PCF). The interworking between HPLMN and VPLMN is protected by home Security Edge Pro-tection Proxy (hSEPP) and visited network's Security proxy (vSEPP).

In this example, the visited network provides functions for the network slice selection (via NSSF), network access control and mobility management (via AMF), data service management (via SMF), and applications (via AF). Involving so many key

45

components, this example serves well to summarize the 5G network functions as presented in the following list and Figure 8.

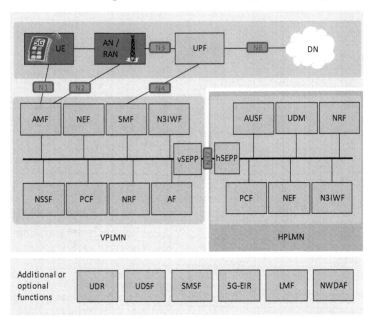

Figure 8 The main functional elements of 5G.

5G-EIR: The optional *5G Equipment Identity Register* (5G-EIR) is an evolved version of LTE EIR. The task of the 5G-EIR is to check the status of the Permanent Equipment Identity (PEI) in case it has been blacklisted. The PEI is an evolved variant of the traditional IMEI, International Mobile Equipment Identity used in the previous mobile networks.

AF: The *Application Function* is comparable with the LTE AS (Application Server) and the GSM Service Control Function (gsmSCF). It interacts within the core network to provide services such as application influence on traffic routing, access to the Network Exposure Function (NEF), and interaction with the policy framework.

AMF: The 5G *Access and Mobility Management Function* replaces the LTE Mobility Management Entity (MME). It takes care of the 5G signaling. The AMF has many tasks such as access authentication and authorization, and transport for Short Message Service (SMS). It embeds the *Security Anchor Functionality* (SEAF) and includes location services management for regulatory services.

AUSF: The 5G *Authentication Server Function* replaces the LTE MME (Mobility Management Entity) and AAA (Authentication, Authorization and Accounting). It supports authentication for the 3GPP access and untrusted non-3GPP access.

LMF: The *Location Management Function* determinates the location of the UE. It provides geodetic location determination for a target UE. Please refer to Location-Based Service Chapter for more information on the LMF.

N3IWF: The *Non-3GPP Interworking Function* takes place when untrusted access such as Wi-Fi is used via 5G infrastructure.

NEF: The 5G *Network Exposure Function* is an evolution of the SCEF (Service Capability Exposure Function) and API layer of LTE. In 5G, it assists in storing and retrieving exposed capabilities and events of the Network Functions (NF) together with the UDR. This information is shared between NFs within 5G network.

NRF: The 5G *Network Function (NF) Repository Function* is a part of the evolution of the Dynamic Name Server (DNS) used in LTE. In 5G, it supports service discovery function for providing the information of the discovered NF instances to the requesting NF instance. One of the tasks of the NRF is to maintain the NF profile and respective services of available NF instances.

NSSF: The 5G *Network Slice Selection Function* supports the 5G-specific network slicing concept. The NSSF selects needed set of Network Slice instances serving the UE.

NWDAF: The *Network Data Analytics Function* is managed by the respective 5G MNO. The NWDAF provides slice-specific network data analytics to the Network Functions which are subscribed to it.

PCF: The 5G *Policy Control Function* is an evolution of the LTE PCRF (Policy and Charging Enforcement Function). In 5G, it supports unified policy framework to govern network behavior. It provides policy rules to the Control Plane functions to enforce them.

SEPP: The 5G *Security Edge Protection Proxy* interconnects 5G networks. The SEPP elements are located at the edge of the 5G core network (the hSEPP residing in home network and the vSEPP in visited network). The SEPP provisions confidentiality and integrity protection of the signaling between networks, and hides the topology of the network behind a firewall. The SEPP is a non-transparent proxy and supports message filtering and policing on the inter-PLMN control plane interfaces.

SMF: The 5G *Session Management Function* replaces, together with the 5G UDF, the LTE Serving Gateway (S-GW) and PDN Gateway (P-GW). Among other tasks, the SMF handles Session Management for session establishment, modification, and release.

SMSF: The *Short Message Service Function* supports the Short Message Service (SMS) over a 5G Non-Access Stratum (NAS). It manages SMS subscription data and takes care of its delivery via signaling channels.

UDM: The *Unified Data Management* generates 3GPP AKA (Authentication and Key Agreement) credentials for each user.

The UDM resides in the same HPLMN with the subscriber it serves. It performs user identification including storage and management of the Subscription Permanent Identifier (SUPI) per each individual 5G subscriber. It also can de-conceal the Subscription Concealed Identifier (SUCI) which is the privacy-protected subscription identifier. Furthermore, the UDM manages access authorizations such as roaming restrictions.

UDR: The 5G *Unified Data Repository* is an evolved version of the LTE SDS (Structured Data Storage). It can store and retrieve subscription data, and it takes care of the storage and retrieval of policy data by the PCF, storage and retrieval of structured data for exposure, and application data by the NEF.

UDSF: The 5G *Unstructured Data Storage Function* is comparable with the LTE SDSF (Structured Data Storage Function). In 5G, the UDSF is an optional function for storing and retrieval of information in a form of unstructured data by any NF.

UPF: The 5G *User Plane Function* replaces, together with the SMF, the LTE S-GW (Serving Gateway) and P-GW (PDN Gateway). It takes care of the user data. The UPF acts as an anchor point for intra- and inter-RAT (Radio Access Technology) mobility. It also is the external PDU session point to interconnect Data Network (DN) and it takes care of packet routing and forwarding.

The 5G core network architecture is predominately service-based, and it can take advantage of the SDN and NFV concepts. The 5G core network is also based on *Network Slicing* which optimizes the offered service type per vertical.

The *Software Defined Networking* (SDN) of 5G refers to a network architecture model that minimizes the hardware constraints by abstraction of the low-level functions. As a consequence, it is capable of executing these functions in a software-based, centralized control plane via the Application Programming Interfaces (API). It also makes the network services agnostic to the underlaying hardware.

Together with the SDN concept, the *Network Functions Virtualization* (NFV) has an important role in maximizing the 5G performance. The principal task of the NFV is to decouple the software from the hardware by virtual machines (VM).

Network Slicing can be understood as "networks over network". With it, a network operator can create a single network, or in typical cases, several parallel virtual networks within a physical network in order to serve a variety of very different types of verticals. As an example, a drone controlling application benefits from ultra-low latency values to ensure the fastest possible response to the remote commands. Meanwhile, an intelligent, permanently installed wireless remote sensor may need only a most basic and occasional data transfer service for a telemetric purpose.

By setting up separate network slices, MNO can optimize the network resources and provide fluent user experience for all the customers and devices.

Because of the much more dynamic operations of the network slice -enabled network compared to the traditional models, the 5G core network needs to manage the slices efficiently and in real time. This takes place via the Network Slice Selection Function.

Further reading	
▶3GPP	*TS 23.501* (system architecture)
▶5G Explained	Section 6.3 (5G core network elements)

Deployment

The 3GPP Release 15 Technical Specifications (TS) were finalized in the beginning of 2019, and many operators started to deploy the first globally interoperable 5G networks soon after, during 2019.

Prior to these global standards, there were early pilots and proprietary 5G networks deployed based mostly on fixed wireless access (FWA) and Customer Premises Equipment (CPE) which refers to a non-mobile user equipment placed at a designated location such as home or office.

South Korea was one of the first countries deploying commercial mobile 5G. According to the Ministry of Science and ICT, the country's three telecoms had a combined 260,000 5G subscribers in the beginning of May 2019. (GSMA, 2019)

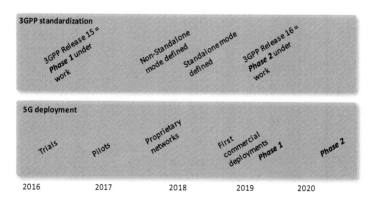

Figure 9 Standardization and deployment of 5G.

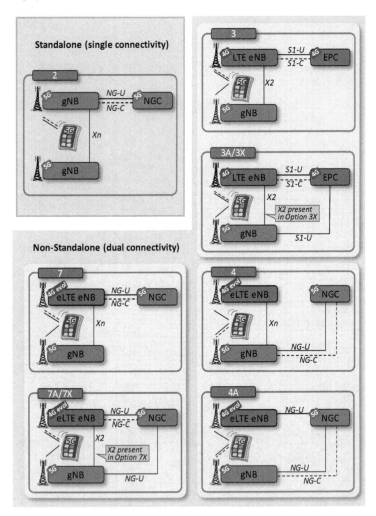

Figure 10 The most typical options for 5G deployment. The dotted line represents signaling (control plane) and the solid line means data connectivity (user plane).

The United States of America will experience one of the fastest customer migrations to 5G in the world. 5G will reach 100 million mobile connections in early 2023, and by the end of 2025, the USA will have more than 190 million 5G connections, meaning that half of the population will be covered by then. (GSMA, 2018) According to GSMA, global 5G implementations are happening more peacefully, with an estimated 15% base reached by 2025.

The 3GPP Release 15 specifications provide the MNOs with a variety of options to deploy 5G in a gradual manner. Following the standard terminology, a full and solely 5G-enabled network refers to the Option 2 as depicted in Figure 10.

The 5G-specific base station is called gNB (next generation Node B). It communicates with the 5G User Equipment (UE) which is user's mobile phone or other 5G device such as sensor.

Prior to the Option 2 deployment, operators may construct the 5G capacity and coverage by relying on the previous 4G infrastructure as described in the Non-Standalone connectivity scenarios. 3GPP has designed many possibilities from which the options 3, 4 and 7 can be assumed to be the most popular ones.

Option 2: This Option represents the Standalone 5G which is the ultimate goal of the 5G operators. It consists only of high-performing 5G radio and core systems.

Option 3: The 5G UE relies on both 4G and 5G radio networks, and solely on 4G core. The 4G core is, in fact, not even aware of the new 5G radio signaling interface in this scenario, and the old LTE base stations (eNB, evolved NodeB) serve as anchors to interconnect the 5G gNB elements and 4G core. The user data can be delivered between the core and only LTE eNB, or via both eNB and 5G gNB. In the Option 3A, there is no *X2* interface. This Option has also variant called 3X which is marked in Figure 10 with the *X2* interface. In the Option 3X, the traffic flow is converged at the 5G gNB and divided from there to the 4G eNB while the 5G NR takes care of the majority of the traffic. Whenever 5G NR has lower coverage performance, the traffic split mechanism can offload more traffic to the eNB. This facilitates the optimized bandwidth in the *X2* interface. The Option 3X is thus highly practical candidate for the initial 5G deployments.

Option 4: The evolution of 5G can lead into an adjusted, intermediate step of the Non-Standalone, NR-assisted architecture that relies on the 5G NGC (Next Generation Core). In this Option, the 5G gNB acts now as the anchor for connecting the data directly and from the further evolved 4G eNB elements (eLTE eNB) between the 5G core.

Option 7: Complexity-wise, this scenario is between the Options 3 and 7. It is similar to the Option 3, but the 4G eNB elements are upgraded to better cope with the delivering of the 5G performance; hence the updated term of the eLTE eNB (evolved LTE eNB).

These examples show that there are many ways the MNOs may deploy their 5G networks in the initial and more mature phases. In the practical environment, the MNOs may consider also the following deployment scenarios as interpreted from the *5G Deployment Considerations* document of Ericsson. (Ericsson, n.d.)

Collocated, same frequencies: Non-Standalone 5G network, 4G and 5G radio using low-band and/or mid-band set in sub-6 GHz. The coverage areas of the 4G and 5G are comparable ensuring seamless user experience whenever the terminal performs handover between 4G and 5G. In this scenario, 5G provides high capacity in dense city centers and urban areas especially for the eMBB and Fixed Wireless Access users, whereas 4G can be used as a "gap filler" for the rest of the areas to serve the less demanding applications and IoT devices. In this scenario, the radio equipment of 4G and 5G may be collocated into the same sites, making it straightforward to reuse antenna towers, transmission, power supplies and site shelters.

Partially collocated, different frequencies: In this scenario, 4G may provide a basic coverage layer on the sub-6 GHz bands while Non-Standalone 5G serves customers on the mm-Wave bands. The customers of the latter case benefit from the highest capacity and data rates while enjoying the lowest possible latency. In this scenario, the physical 4G sites can be reused as such for 5G, but the denser 5G gNB network requires also additional sites. This scenario is feasible for also enhancing the already existing 4G network by collocating 4G radio equipment to the new 5G sites.

5G-focused: In this scenario, Standalone 5G, as defined in the Option 2, operates on all the possible bands of the operator including low, mid and high-bands. The low-bands provide large coverage for the basic 5G services while the mid-band is useful especially in the urban areas, balancing the radio coverage and data speeds. The mm-Wave provides the users with the highest data speeds for the most demanding data services especially in the dense urban environment.

The deployment of 5G will be gradual, and as the network coverage enhances over the years, there will be increasing number of diverse devices capable of taking fully advantage of the New Radio of 5G. The year 2019 will be the time for Non-Standalone deployments, until the Release 16 and gradual Standalone networks start facilitating the full potential of 5G.

The high-level 5G scenarios can be thus categorized into Non-Standalone and Standalone deployments which are described more detailed in the Annex J of the 3GPP Technical Report TR 23.799. Furthermore, the 3GPP TR 38.801 presents practical deployment scenarios summarized in the following.

Non-centralized deployment: This scenario refers to a set of 5G gNB elements equipped with a full protocol stack. The scenario is suitable especially in macro cell and indoor hotspot deployments. The gNB elements can be connected with other gNB elements or LTE eNB and eLTE eNB elements.

Co-sited with E-UTRA: In this scenario, the 5G NR functionality is co-sited with the 4G E-UTRA. This deployment scenario is suitable for many cell types, including the Urban Macro which provides the largest coverage. A load balancing can be used in this scenario to optimize the 4G and 5G radio access spectrum resources.

Centralized deployment: The 5G NR supports centralization of the upper layers of the NR radio protocols, and various gNB elements can be attached into the centralized unit via a transport network. The non-collocated and collocated deployment scenarios with the 4G E-UTRA are applicable in this scenario.

Shared RAN deployment: This scenario refers to an environment where various hosted core operators are present.

Heterogenous deployment: This scenario refers to the heterogenous RAN service areas. It can be an optimal solution for the indoor deployments in order to ensure a fluent interworking between shared RAN and non-shared RAN.

Further reading	
▶3GPP	*TS 38.300* (5G new radio)
▶5G Explained	Section 4.2.3 (gradual deployment of 5G)
▶GSMA	*Road to 5G: Introduction and Migration*

Edge Computing

5G relies increasingly on the cloud concept. Clouds enable intelligent service awareness which, in turn, optimizes connectivity, latency, and other performance characteristics. 5G networks are also highly scalable and they offer advanced services thanks to the virtualization of the network functions.

One of the consequences of this evolution is that 5G will rely largely on data centers. The 5G ecosystem has centralized, regional and edge data centers as depicted in Figure 11.

In addition to centralized clouds residing in the main data centers, 5G applications may be located at the network's edge, closer to the end-user. These applications can be hosted in the mobile edge computing nodes, referred to as cloudlets.

5G requires thus high-capacity cloud processing power to fulfill its demanding performance criteria. As 5G deployments evolve, there will be need for further data storage, increased server capacity, additional cooling for the equipment, and more space for housing the respective racks.

Oftentimes, the same data center will serve 5G as well as many other systems. The preparedness can be seen already in practice as data centers are being deployed actively over wide areas.

Data centers will be changing from the previous decentralized mobile network model to better serve the centralized processing of the 5G network functionalities. This means that instead of using stand-alone equipment for each function of the network, 5G will rely on cloud-based solutions both in main data centers and edge regions.

The 5G network functions run thus on virtualized software environments instead of standalone HW elements. This principle could potentially evolve further so that major part of the base station processing can take actually place in the cloud, whereas the base station could occasionally process more, e.g., when cloud's own load threshold reaches a predefined limit.

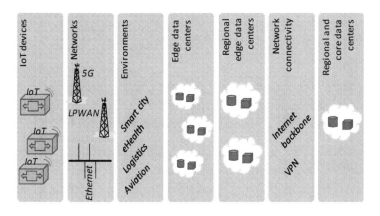

Figure 11 Cloud ecosystem with edge data centers.

An edge cloud can bring both the contents and data processing closer to the end-users. One of the clearest benefits of the edge cloud is the reduced latency as depicted in Figure 12. The closer to the mobile device the contents can be located, the lower the latency, thanks to the shorter transmission path. The edge can be located either in cloud core network or cloud RAN, or even at the base station site.

Another benefit of the cloud edge is that part of the data processing of the device such as smart phone can be offloaded to it. 5G will make this scenario reasonable in many cases thanks to the high data speeds and low latency.

Practical example of this is an augmented reality (AR) / virtual reality (VR) device which typically requires heavy processing power in order to provide fluent user experience.

5G connectivity could be used for the AR/VR device embedding the device to a headset display. The challenge is that the mobile phone is not capable of processing as efficiently data as the standalone hardware is. To overcome this issue, the processing of the contents could be offloaded into the cloud edge while the 5G device serves the user merely as a relatively simple interface between the display and 5G connectivity which is able to support high-speed, low-latency transmission modes.

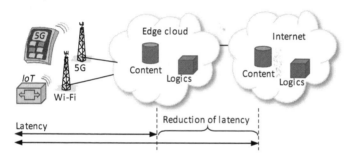

Figure 12 The edge cloud reduces latency.

The increasing number of such demanding applications could take advantage of the 5G connectivity and cloud computing. At the same time, the terminal cost could be optimized because it would not need so much processing power, as long as it supports the eMBB and URLLC connectivity to the edge cloud.

Business-wise, edge computing and data centers open new opportunities for both established and new stakeholders. Some operators may want to deploy and manage their edge infrastructure, but it could be equally feasible to buy or lease the cloud processing and capacity based on Cloud-as-a-Service model.

The virtualization of the 5G architecture opens this opportunity on a completely new level compared to earlier generations. The older networks are more isolated environments where operators take care of their own stand-alone network components (machines with dedicated HW and SW running on top) whereas the functions are instance-based and HW-agnostic in 5G.

The infrastructure model can also be of hybrid consisting own and purchased services depending on the deployment phase, area of operation, and other factors. This mode makes it possible to optimize the Return of Investment (RoI). Especially for the smaller 5G operators, it might not make sense to build a large data center at least in the beginning of the deployment.

Outsourcing the cloud-based 5G Network Functions and other tasks generates new business models. There might also be a need to take into account the assurance of the service availability. There is typically an SLA (Service Level Agreement) negotiated between an operator and cloud service provider to set the expectations for the service up-time. In the critical environment, it can be so called "five nines", i.e., the service is guaranteed to stay up 99.999% of the time. That may require an active-active pairing of parallel servers in geo-redundant configuration which provides highest assurance level, but with an increased cost. For the less critical applications, less expensive passive-passive configuration and non-georedundancy may be an adequate solution.

Further reading	
▶ITU	*ITU-T Y.3515*
▶5G Explained	Section 6.2.1 (Cloud concept)

Frequencies

To cope with the new requirements, 5G networks support extended radio frequency bands and bandwidths. There will be many new bands below and above 6 GHz to help operators to cope with the increased capacity demand.

The globally agreed frequencies will be discussed at the WRC-19 of ITU-R, whereas the country-specific deployments depend on the regional regulators. The current discussions include many possible variants up to around 100 GHz bands. The most feasible frequency strategies favor wide and contiguous bands.

The ITU-R WRC-15 identified a set of frequencies to be studied for the the 5G, including 24.25–27.50 GHz, 37.00–40.50 GHz, 42.50–43.50 GHz, 45.50–47.00 GHz, 47.20–50.20 GHz, 50.40–52.60 GHz, 66.00–76.00 GHz and 81.00–86.00 GHz. These frequencies are being assessed for the use on a primary basis. There is already a set of 5G radio frequency bands approved by 3GPP as summarized in Figure 13.

There are also many frequency bands under evaluation requiring possibly additional allocations to the mobile service on a primary basis, and these bands include 31.80–33.40 GHz, 40.50–42.50 GHz and 47.00–47.20 GHz.

In practice, the first frequencies have already been auctioned for the initial 5G deployments during the first half of 2019. As an example, US operators have purchased capacity on 2.4, 3.5, 24 and 28 GHz bands. There are also other regulators that are investigating the options for the mobile industry's preferred bands above 30 GHz.

5G NR supports different UE channel bandwidths in a flexible way. The base station can transmit to and receive from one or more UE bandwidth sections in any part of the carrier resource blocks. Along with the results of the ongoing feasibility studies for the 5G frequency bands, the currently concrete bands are listed in 3GPP TS 38.104. It specifies transmission bandwidth configurations for the Frequency Ranges FR1 and FR2.

Table 4 5G Frequency Range definitions.

Frequency Range	MHz	Type
FR1	450 – 6,000	Sub-6 GHz
FR2	24,250 – 52,600	mm-Wave

The **FR1** transmission bandwidth configurations can have bandwidth values of 5, 10, 15, 20, 25, 30, 40, 50, 60, 70, 80, 90 and 100 MHz whereas the subcarriers can be varied between the values of 15, 30 and 60 kHz. For the **FR2** mode, the transmission bandwidth configuration can have bandwidth values of 50, 100, 200 and 400 MHz, and the subcarriers can be either 60 or 120 kHz.

Figure 13 summarizes the identified 5G frequency bands as per the 3GPP TS 38.104 dated in July 2019. There will be additional frequencies in the future versions of this specification. Currently, there are 33 FR1 bands listed, and 4 FR2 bands for the mm-Wave area.

As can be noted, the number of mm-Wave bands is still limited at this time. We may expect additional bands to be agreed along with the Release 16 as the work progresses in 3GPP, and as soon as the new frequency band allocations will be decided at the WTC-19. The new versions of the 3GPP TS 38.104 will reflect these new bands as they become concrete.

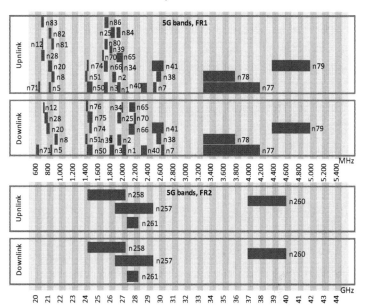

Figure 13 The 5G NR bands as defined in 3GPP Release 15 (status on July 2019).

5G uses FDD (Frequency Division Duplex) and TDD (Time Division Duplex) frequency bands. 5G introduces also a SUL (Supplementary Uplink) concept. Figure 14 depicts the principle of SUL, and it is detailed in 3GPP TS 38.300.

The idea of SUL in 5G is to compensate the cell radio coverage in the uplink direction. The user equipment's transmitted power level in the uplink is typically very low compared to the 5G gNB power levels. This may lead in a considerable performance degradation closer to the cell edge area causing unbalance between the uplink and downlink directions. This can be seen, e.g., by failing two-way mobile video conferencing, especially if there is no sufficient overlapping in the cell edge area.

As the lower frequencies propagate wider distances, the idea is to use such secondary supplementary frequency band to enhance the uplink performance of the user equipment. In 5G, the concept is adaptive, so the default uplink band is used as such in normal condition. As soon as the quality decreases below a defined threshold value on a cell edge area, 5G network orders the user equipment to use a better-propagating SUL band for its uplink transmission instead.

The currently defined 5G SUL bands are n81, n82 and n83 which are located at the sub-1 GHz spectrum to ensure large coverage area. The bands n80, n84 and n86 all are below the 2 GHz spectrum which still provides adequate coverage.

In the practical deployments, the potential interference with the overlapping LTE frequencies needs to be researched case-by-case based on 3GPP TR 37.872 and TR 37.716.

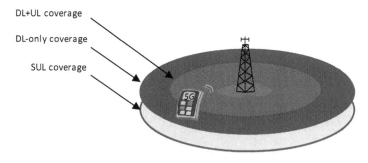

DL+UL coverage

DL-only coverage

SUL coverage

Figure 14 The principle of 5G Supplementary Band.

Further reading	
▶3GPP	*TS 38.104* (5G base station radio transmission)
▶5G Explained	Section 5.3 (5G spectrum)

Future (where's the 6G?)

5G is already reality in the commercial markets. As has been a "tradition" since the first generation of mobile communications, a new generation is introduced for each new decade. So, one might wonder what is the status of the forthcoming generation?

The next step beyond the "visible horizon" could be the sixth generation of mobile communications, 6G. Before it will become even a standardization item, there is plenty of work to do in the 5G era. The first phase of 5G is about to take off during 2019, and 5G with its forthcoming releases will serve us for years to come. The network deployment of the 3GPP Release 16-based second phase 5G begins during the better half of 2020. After that, 5G will evolve through several 3GPP releases, similarly as we have already seen with the previous systems; 3GPP is already collecting ideas for the Release 17 and beyond.

While 5G starts taking over and will serve us throughout 2020s, ITU is already foreseeing the future needs. The timing coincides with the familiar cyclic pattern; a new generation has been deployed as soon as the new decade begins. ITU has recently started discussions on how the mobile communications could look like in 2030. The group dealing with this topic is called ITU-T Focus Group Technologies for Network 2030, FG NET-2030. (ITU, 2018)

The FG NET-2030 was established by the ITU-T Study Group number 13 in 2018, and its aim is to investigate expected capabilities of networks for the year 2030 and beyond. The study includes future network architecture, requirements, use cases, and capabilities of the networks.

The task is interesting as there are already many scenarios foreseen that might be challenging to cope with by evolved 5G, including holographic type communications, ultra-fast critical communications, and high-precision communication demands of emerging market verticals. The task of the group is to figure out the most suitable, potential network architectures and respective mechanisms for the identified communication scenarios, and document the findings in a form of Network 2030.

The group is not restricted to existing solutions but their evolution is part of the research. Although there is not too much concrete information available on the expected outcome quite yet, one of the basic principles of the Network 2030 system will be to ensure backwards compatibility so that the networks can support both existing and new applications.

In addition to the above-mentioned gap and challenge analysis, the next step of the FG NET-2030 includes the collection of all aspects of Network 2030, including vision, requirements, architecture, novel use cases, and evaluation methodology. As a part of the activities, the group will provide guidelines for standardization roadmap and establishes liaisons with other standardization development organizations.

For the interested ones, participation in the FG NET-2030 is open to all. More information on the initiative can be found at the web page of the group. (ITU, 2018)

Further reading	
▶ITU	*FG NET-2030* (ITU, 2018)

Generations

Mobile communication systems develop fast, and the trend has been to get a completely new generation up and running each decade since the initial 1G in 80s. It's thus time for a new era again – the theme being about connected societies in 5G now.

Increasing number of countries will have launched initial 5G networks by the end of 2019. The year 2019 is important for the 5G smartphone launches, and the World Radiocommunication Conference 2019 (WRC-19) will add and align 5G frequency bands for the further optimization of the radio.

We have been able to use commercial mobile communication networks since 1980s. The early systems at that time were of 1st generation, and they offered mainly voice service via analogue channels. (Penttinen J. , Mobile Generations Explained, 2015)

1G refers to analogue, automatic mobile networks that were meant for only voice calls, although accessory-based solutions were sometimes possible for the data transfer. The initial systems used vehicle mounted and portable devices. The weight of such device was typically several kilograms. Some examples of this 1st phase of 1G were NMT-450, Netz-C, and AMPS.

As 1G matured, also hand-held devices became popular. The first ones were big and heavy compared to modern devices – those were definitely not meant for pockets. Example of this latter phase is the NMT-900 system which was launched in Nordic countries 1986-87.

2G represents digital systems which integrate data services. Examples of this generation are GSM and IS-95. GSM was launched commercially 1991, and unlike other 2G variants at that time, it is based on SIM (Subscriber Identity Module) for housing subscription related data.

SIM has evolved ever since. It is still useful platform for storing user's unique key which is the base for authentication and authorization of the user, and serves also for radio interface encryption. It is a hardware-based secure element (SE). 5G will rely on it, too, in one or another form.

The 2G data speeds were originally as low as 9.6 kb/s, and the service was based on circuit switched connectivity. GPRS (General Packet Radio Service) was deployed in 2000, and it started to offer packet switched IP data. The data speed has increased along with the further evolution of GSM. Using multislot and multicarrier technologies, the speed can nowadays be over 1 Mb/s depending on the service support on network and device.

Due to the low spectral efficiency and security, the importance of 2G is lowering and its frequencies are being re-farmed for other systems. Nevertheless, 2G is still used in many markets for consumer and Machine-to-Machine (M2M) communications such as wireless alarm systems, so only time will show when the 2G sunset will occur at larger scale.

3G is a result of the further development of multimedia-capable systems which provide much faster data speeds. 3G can be thus characterized as a mobile multimedia platform. ITU's IMT-2000 (International Mobile Telecommunications) sets the performance requirements for 3G systems. There are various commercial 3G systems such as US-originated cdma2000 and 3GPP-based UMTS/HSPA. 3G networks have evolved since their commercial launch in the beginning of 2000, and today, they are capable of supporting tens of Mb/s data speeds.

4G continued with the "tradition" of renewed generations. ITU-R designed a set of IMT-Advanced requirements for the 4G systems. There are two commercial systems complying with them; LTE-Advanced (LTE-A) specified as of 3GPP Release 10, and WiMAX (WirelessMAN-Advanced) which is based on the IEEE 802.16 evolution. Oftentimes in commercial field, the LTE as per Release 8 and 9 is considered to belong to 4G, and there have been operators interpreting even HSPA+ to be part of 4G. Nevertheless, referring strictly to the IMT-Advanced, they are merely 3G technologies. Nowadays, the significance of the WiMAX has decreased considerably, leaving LTE-Advanced as the only relevant representative of 4G era. 4G offers nowadays hundreds of Mb/s data speeds.

5G refers to the systems beyond IMT-Advanced which comply with the new ITU IMT-2020 requirements. 5G provides again much higher data speeds. The full version of 5G is expected to be available for the commercial use as of the second half of 2020. The initial 5G, as defined in 3GPP Release 15, is oftentimes called *phase 1* 5G whereas the Release 16 represents *phase 2* 5G. The latter will comply with the strict requirements of the IMT-2020, making it a complete 5G which provides the customers with the full set of services and highest performance.

As new generations take over and customers start enjoying their enhanced and much more spectrum-efficient performance, previous generations gradually lose their users. Although 2G and 3G still have important use base, including IoT devices, operators may consider already the decommissioning strategies.

Meanwhile, it may be assumed that many operators re-farm 2G and possibly 3G bands to 4G and 5G to optimize the use of spectrum. In this transition phase, the already existing base station sites and possibly part of their equipment can be reused, including power supplies and transport lines.

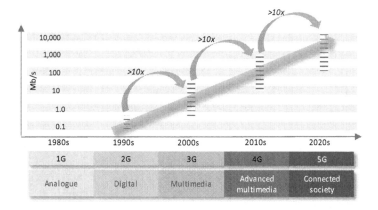

Figure 15 Mobile generations vs. downlink data speed evolution.

Each generation lives throughout series of enhancements. Since the very initial deployments of 2G, the data services of each generation have continued evolving, providing the users with constantly enhancing performance and capacity. As depicted in Figure 15, this phenomenon can be generalized by stating that each new generation provides at least 10-fold data speed ranges for the customers compared to the previous generation.

5G is of no exception, so, while LTE-Advanced is capable of delivering some hundreds of Mb/s up to about 1 Gb/s range, the eMBB mode of 5G can be assumed to handle downlink data speeds of around 10-20 Gb/s.

Further reading	
▶3GPP	*TS 22.261* (service requirements for next generation new services and markets)
▶5G Explained	Section 3.2 (mobile generations)

Health Considerations

Along with the introduction of cellular communication systems, since their initial commercial deployment, there have been questions about potential health risks of the radio waves. 5G is of no exception.

5G and all the previous generations of mobile communication systems are based on radio wave propagation over the air. In 5G, the respective RF (Radio Frequency) spectrum varies from few hundred-megahertz (MHz) up to tens of gigahertz (GHz) frequencies as summarized in Radio Frequencies Chapter.

Radio waves have formed a foundation for our communication services for decades, including wireless telegraphy, radio and TV, walkie-talkies, radars, navigation and Wi-Fi. Since the introduction of the first generation mobile communication networks in 1980s, the popularity of the voice and data communications on-the-go has increased considerably. According to the ITU and FCC statistics, there were 391.6 Million mobile device users in the USA in 2017 whereas the number was merely 109.5 Million in 2000. (ITU, 2019)

Every now and then, questions emerge on the potential health issues of the radio communications. As the radio waves cannot be observed by the human eye, the topic may be especially susceptible for also somewhat belief-based statements. Many research organizations have thus put considerable efforts to investigate the topic and to find clarity to the doubts.

Intuitively thinking, it is tempting to associate the use of a mobile phone, e.g., with brain tumors as the devices are oftentimes used close to the head. Nevertheless, as there have been similar health issues also prior to the mobile communications took off, this reasoning needs to be verified in order to understand if the diagnosed decease and use of the mobile phone correlate, or if the root cause may relate to something else such as other type of cancer which spreads into the brain.

For the ones interested in exploring the topic, it is recommendable to verify the latest information from the reports of recognized expert organizations as they rely on objective and scientific research methodologies. The aim of the scientific studies is to reason whether the investigated item, such as the use of mobile phone, can be shown to contribute systematically to brain tumors or other health issues while assessing other potential sources that may contribute to the outcome.

The quality of the research may logically vary. To understand the relevance of the study, it should be possible to obtain similar results by any other independent research applying the same, documented and peer-reviewed methodology. If there are contradicting results, it is not possible to agree on the causality; instead, it may indicate even weaknesses in the study setup.

The challenge is that there may be various harmful elements in the environment we live in, and in products we consume, and which may unwillingly contribute to the study results. As an example, the research could be impacted by previous or current exposure on, e.g., radon gas, radiation treatment, asbestos and other environmental carcinogens, to mention some. Also, naturally occurring diseases, exposure to ultra violet radiation of the sun and environmental chemicals, and many more reasons may interfere the analysis. As stated in (NIH, 2019), there are many factors influencing whether a person exposed to a carcinogen will develop cancer.

Another challenge is that there must be enough samples in order to obtain statistically significant results. In the most accurate research, there should be significant amount of study objects in an environmentally isolated location for a long period of time in order to minimize the effect of any other potential causes, exposing part of the objects to RF radiation that complies with the current regulatory safety limits while the rest of the study objects is kept out of the exposure. That scenario is obviously challenging to carry out in practice, so there is a need to understand well the significance of any other interfering factors that may contribute to the final results.

If the study is designed poorly with too many unknowns and assumptions, it is easy to go wrong with the methodology and interpretation of the outcome resulting funding-bias, false-positives and false-negatives.

Ionizing radiation, originated from such sources as X-ray equipment, has enough energy to detach electrons from atoms or molecules ionizing them. As a consequence, the ionizing radiation can harm human cells.

Mobile communication is based on **non-ionizing** radiation so it cannot modify atoms or molecules. Nevertheless, strong RF radiation originated from, e.g., high-power pulse radar can heat living tissue. Another example of this effect is a microwave oven which uses several hundreds of watts on 2.4 GHz band as it coincides with the resonance frequency of water molecules. This resonating makes the molecules warm up due to the friction of the vibrating molecules.

The mobile industry follows thus established rules set by regulators for the RF radiation safety limits. As an example, although Wi-Fi device can work on the same band as microwave oven, it would not cause thermal effects. Mobile communication specifications are designed accordingly limiting the maximum radiating power level of both base stations as well as mobile devices. Operators need to ensure that the total maximum radiating power levels of base stations stay within the regulated range while mobile device manufacturers design the power levels of their equipment to comply with the limits.

The non-ionizing radio frequency radiation can be estimated by technical values from which the typically applied is the SAR (Specific Absorption Rate). The SAR is defined as the amount of energy absorbed by a mass of a biological tissue. The SAR is a unique, measured value for every model of mobile phone. In many countries, the regulators dictate the limits of the respective values, the maximum value being typically in range of 1.6 – 2 W/kg. As an example, FCC has set the limit to 1.6 W/kg in the USA. (FCC, 2016)

The power levels of the mobile communication base station equipment are a small fragment of radar systems or microwave ovens, and in addition, the RF power lowers exponentially as a function of the distance from the antenna. The heat effect due to normal mobile device's RF radiation may be assumed to be so insignificant it is challenging to even measure the increased heat from living tissue. For special cases such as the installation personnel climbing up to an antenna tower, there are regulated safety zone distances for the close proximity of the base station antennas which depend on their radiating pattern and power.

There have been number of studies seeking for indications of potential impacts of the RF radiation. (FCC, 2015) Some programs have been established for a long term and involve large number of people. In addition to laboratory research and studies in people, some studies have researched the overall increase of the mobile device use and compared it with the generic statistics of the number of cancer patients. Observing the references presented in this chapter, it seems that up to day, the scientifically adequate study reports have not found a systematic or alarming relation between human health issues and typical use of the cellular devices. Nevertheless, as part of the published studies indicate mixed results, the US National Cancer Institute (National Cancer Institute, 2019) cites the following:

"In 2011, the American Cancer Society (ACS) stated that the IARC classification means that there could be some cancer risk associated with radiofrequency radiation, but the evidence is not strong enough to be considered causal and needs to be investigated further. Individuals who are concerned about radiofrequency radiation exposure can limit their exposure, including using an ear piece and limiting cell phone use, particularly among children."

The US National Cancer Institute has also collected statements of relevant agencies. The Institute states the following:

"In summary, most studies of people published so far have not found a link between cell phone use and the development of tumors. However, these studies have had some important limitations that make them unlikely to end the controversy about whether cell phone use affects cancer risk." ... "With these limitations in mind, it is important that the possible risk of cell phone exposure continue to be researched using strong study methods, especially with regard to use by children and longer-term use."

The publication of the Minister of Public Works and Government Services "Limits of Human Exposure to Radiofrequency Electromagnetic Fields in the Frequency Range from 3 kHz to 399 GHz; Safety Code 6" summarizes below study reports it lists in their web page. (Health Canada, 2018)

"*Despite the advent of numerous additional research studies on RF fields and health, the only established adverse health effects associated with RF field exposures in the frequency range from 3 kHz to 300 GHz relate to the occurrence of tissue heating and nerve stimulation (NS) from short-term (acute) exposures. At present, there is no scientific basis for the occurrence of acute, chronic and/or cumulative adverse health risks from RF field exposure at levels below the limits outlined in Safety Code 6. The hypotheses of other proposed adverse health effects occurring at levels below the exposure limits outlined in Safety Code 6 suffer from a lack of evidence of causality, biological plausibility and reproducibility and do not provide a credible foundation for making science-based recommendations for limiting human exposures to low-intensity RF fields.*"

Along with the introduction of 5G, there seem to be also some-what confusing messages, or even disinformation, circulating in the Internet. Swisscom is one of the entities collecting state-ments on the mobile safety topics in order to provide answers. (Swisscom, 2019) Their web presents, e.g., the following:

"*According to all current scientific knowledge, there are no health risks that can be attributed to mobile radiation. When it comes to cancer, however, the data available is still somewhat unclear. Thus, to be on the safe side, the International Association for Research on Cancer IARC classifies mobile phone radiation as "possibly carcinogenic" – the exact same classification they give to coffee and many other substances.*"

As indicated in Ref. (GSMA, 2019), expert groups and public health agencies such as the World Health Organization broadly agree that no health risks have been established from exposure to the low-level radio signals used for mobile communications.

Referring further to the source (GSMA, 2019), based on expe-rience with 3G and 4G networks and the results from 5G trials, the overall levels in the community will remain well below the international safety guidelines. Also, compliance assessment of 5G network antennas and devices are dictated by international standards which include new approaches for smart antennas and the use of new frequency ranges.

The references (WHO, 2019), (FCC, 2015), (National Cancer Institute, 2019), (Health Canada, 2018) and (Swisscom, 2019) can help the interested ones in exploring a variety of studies.

Further reading	
▶World Health Organization	*Establishing a dialogue on risks from electromagnetic fields.* (WHO, 2019)
▶US National Cancer Institute	*Cellular phones and cancer risk* (National Cancer Institute, 2019)
▶Health Canada	*Safety Code 6, 2015* (Health Canada, 2018)
▶Swisscom	*5G Mobile Technology Fact Check* (Swisscom, 2019)
▶GSMA	*Safety of 5G Mobile Networks* (GSMA, 2019)
▶The Telecommunications Handbook	Chapter 24 (Penttinen J. , EMF - Radiation Safety and Health Aspects, 2015)

Identifiers

5G system uses renewed identifiers in order to provide unique means to recognize subscriptions and network elements in the new network ecosystem. Also, new methods are applied for the identifiers such as concealing and PKI (Public Key Infrastructure).

Of the new 5G identifiers, the most important ones are the SUPI (Subscription Permanent Identifier), SUCI (Subscription Concealed Identifier), GPSI (Generic Public Subscription Identifier), GUAMI (Globally Unique AMF Identifier), and PEI (Permanent Equipment Identifier). Also, the established identifiers are valid for the interoperability, including MSISDN (Mobile Subscriber ISDN number), IMSI (International Mobile Subscriber Identity), IMEI (International Mobile Equipment Identity), and NAI (Network Access Identifier).

For the mapping of the identifiers in different scenarios, please refer to the 3GPP TS 29.571 which details the 5G data types for the subscription, identification and numbering.

SUPI refers to a *Subscription Permanent Identifier*. It is a primary identifier in 5G, and it forms the foundation for all the key derivation scenarios together with the subscribers' unique K key. The serving network authenticates the SUPI during the authentication and key agreement procedures between the UE and network. Afterwards, the serving network authorizes the UE to use services through the subscription profile obtained from the home network.

SUCI (Subscription Concealed Identifier) is the concealed version of the SUPI. As defined in the 3GPP TS 33.501, the SUCI is a one-time subscription identifier which contains a concealed subscription identifier such as MSIN (Mobile Subscription Identification Number). The network forms a new SUCI as per need.

The SUCI is an optional identifier which is managed by the UICC (Universal Integrated Circuit Card), that is, the 5G SIM card (Subscriber Identity Module). The SUCI provides additional security as it hides the permanent user identification.

The UE generates the SUCI using a protection scheme. It is based on a public key that the user's home network has provisioned beforehand in a secure manner. Based on the indication of USIM, and dictated by the MNO, the forming of the SUCI can be done either by the USIM or the ME. After that, the UE forms a scheme-input for the subscription identifier of the SUPI and executes the protection scheme.

The UE does not conceal the home network identifiers, Mobile Country Code (MCC) and Mobile Network Code (MNC). This is because they are needed for the home network routing and protection scheme.

Please note that there is no requirement for protecting the SUPI in case of unauthenticated emergency call.

Further reading	
▶3GPP	*TS 29.571* (common data types); *TS 33.501* (5G security architecture)
▶5G Explained	Chapter 8.9.3 (5G identifiers)

Interfaces

Along with the renewed physical network elements and a variety of mandatory and optional network functions, there are also many new interfaces in 5G.

The virtualized architecture model of 5G differs from the legacy systems, and there are also plenty of new interfaces.

The key 5G radio network interfaces are listed in Table 5, and Table 6 summarizes some of the most important 5G core network interfaces. The network functions mentioned in Table 6 are summarized more detailed in Core Chapter.

Table 5 5G NR-RAN interfaces.

Interface	Description	Source
NG	NG refers to New Radio which is renewed radio interface between the UE and gNB. It is based on OFDM both in downlink and uplink.	TS 38.410...38.414, NG interface general aspects and principles
Xn / X2	Xn interface is between NG-RAN nodes gNB and ng-eNB. The interface between LTE eNB elements is X2.	TS 38.420...38.424, Xn interface general aspects and principles
E1	E1 is point-to-point interface between gNB-CU-CP and a gNB-CU-UP.	TS 38.460...38.463
F1	Interface between gNB-CU (Central Unit) and gNB-Distributed Unit (DU) elements.	TS 38.470...38.475, F1 interface general aspects and principles
F2	Interface between lower and upper parts of the 5G NR physical layer; includes F2-C and F2-U.	TS 38.300, NR Overall description

Table 6 The most important 5G core network interfaces (reference points).

Interface	Element a	Element b	Description
NG1	UE	AMF	User Equipment – Access and Mobility Management Function
NG2	AN/RAN	AMF	(R)AN – Access and Mobility Management Function.
NG3	AN/RAN	UPF	(R)AN – User Plane Function
NG4	SMF	UPF	Session Management Function – User Plane Function
NG5	PCF	AF	Policy Control Function – Application Function
NG6	UPF	DN	User Plane Function – Data Network
NG7	SMF	PCF	Session Management Function –Policy Control Function
NG8	UDM	AMF	Unified Data Management - Access and Mobility Management Function
NG9	UPF	UPF	Two User Plane Functions
NG10	UDM	SMF	Unified Data Management – Session Management Function
NG11	AMF	SMF	Access and Mobility Management Function –Session Management Function
NG12	AMF	AUSF	Access and Mobility Management Function –Authentication Server Function
NG13	UDM	AUSF	Unified Data Management – Authentication Server Function
NG14	AMF	AMF	Access and Mobility Management Functions
NG15	PCF	AMF	a) *Non-roaming scenario*: Policy Control Function – Access and Mobility Management Function; b) *Roaming scenario*: V-PCF – AMF
N27	hNRF	vNRF	NRF in the Home PLMN is known as the hNRF which is referenced by the vNRF via the N27 interface
N32	hSEPP	vSEPP	Roaming interface
NWu	UE	AMF	N3IWF relays the IPsec between UE and AMF

Further reading	
▶3GPP	*TS 38.300* (New Radio); *TS 38.401* (radio access network); *TS 23.501* (system architecture);
▶5G Explained	Section 6.6 (protocols and interfaces)

Java and APIs in 5G

5G is based on the virtualization of the network functions. It also uses the principles of the Open Source and Software Defined Networking. Practical means for the information transfer between the interfaces is thus needed, as well as hardware-agnostic ways to execute code. Java and APIs offer some of these means.

5G is an end-to-end telecommunication system. It integrates and converges various network types including wireless and fixed system. Furthermore, the 5G radio, core and transport alike rely strongly on cloud infrastructure. The former distributed networks and web application services are not able to comply with the strict 5G networking requirements, but the API and REST concepts can facilitate the new ecosystem to do so.

The role of the **API** (Application Programming Interface) is very important in 5G. Among other indications, the ITU Focus Group IMT-2020 is emphasizing their relevancy in the 5G ecosystem as they facilitate the adaptation of applications and services to deal with the programmable network functions, and help applications to communicate with each other in the highly virtualized mobile communication infrastructure.

One of the key statements of the IMT-2020 is the following:

"Operators of IMT-2020 network should expose network capabilities to 3rd party ISPs (Internet Service Providers) / ICPs (Internet Content Providers) via

open APIs to allow agile service creation, flexible and efficient use of the ca-
pabilities." ... "For IMT-2020, there is an increased need on service custom-
ization by the service providers whereby some of them will offer their cus-
tomers the possibility to customize their own services through service-related
APIs in order to support the creation, provisioning and management of ser-
vices."

APIs are, in fact, essential to interconnect systems and to share data. APIs can help to reduce the cost of the 5G ecosystem because API-interconnected systems can use common software functions and thus optimize the software production.

Another important component in 5G ecosystem is the **REST** (Representational State Transfer). It is an architectural style designed for the distributed hypermedia systems. The REST has become important along with the popularity of the geographic web. The REST works well for sharing information. RESTful Web services integrate to the web as a transport medium. They are less strict for bandwidth, processing power and memory compared to earlier models, and they are capable of communicating through firewalls and proxy web servers. (What is REST, n.d.)

REST is thus an adequate component for 5G in the efforts to interconnect societies. It eases programming and collaboration between stakeholders and promotes multivendor and multi-operator ecosystem.

In a typical 5G network infrastructure, when offering services to the enterprises and end-users, an API Gateway exposes the REST APIs to the 3rd party applications and partners. The API Gateway serves as an entry point which routes requests and is able to do protocol conversion. It is beneficial for cooperating parties such as developers who want to deploy own user interfaces or transfer information via APIs.

Also, the Java has essential role in many 5G functions and procedures. It is a class-based, object-oriented, largely implementation-independent programming language for generic purposes. The aim of the Java is to provide means to design software once, and run the complied Java code on different platforms without need for further adaptations.

JSON (JavaScript Object Notation) is a standardized, language-independent file format based on human-readable text. It is able to deliver data objects that contain attributes with their values, and array data types. As the popularity of the XML format is declining, the JSON is nowadays a popular data format for asynchronous communications between the browser and server. It also is an adequate base for many 5G-related procedures within the 5G system and with external entities.

The mobile communication industry has also used a special version of the Java environment for the SIM card production. Speaking on the UICC (Universal Integrated Circuit Card) and its USIM (Universal Subscriber Identity Module) application, their Card Operating System (COS) is typically card vendor's proprietary solution. Because of such different environments, each of the USIM card applications would need to be adjusted for those different operating systems. That would be a waste of time and resources.

To overcome this issue, JCRE (Java Card Run Time Environment) provides an abstraction layer between each card vendor's own OS variant and the apps running on them as Java applets. As the apps are OS-agnostic in JCRE, the app needs to be developed only once and it is compatible with any OS which supports the abstraction. It is expected that the same principle continues also in 5G era.

JTAPI is relevant for 5G era, too. It refers to a Java Telephony Application Programming Interface. It supports the telephony call control, and it is an extensible API designed to scale and to be used in a range of domains from the first-party call control in a consumer device to a third-party call control in large distributed call centers. (Oracle, n.d.)

Further reading	
▶3GPP	*TS 43.019* (SIM API for Java Card); *TS 31.130* (SIM/USIM API for Java Card)
▶ETSI	*TS 143 019 - V5.6.0* (SIM API for Java Card)
▶ITU	*ITU-T Focus Group IMT-2020 Deliverables* (ITU-T, 2017)

Key Derivation

The security of 5G has been redefined. In fact, only the initial authentication procedure is still the same with the legacy systems whereas the 5G key derivation procedure is completely new.

3GPP has defined a 5G key hierarchy as depicted in Figure 16. The 5G key lengths are by default 128 bits in the initial phase while the network interfaces support key lengths of up to 256 bits as per need for the future purposes. Table 7 summarizes the principle of the 5G keys.

Figure 16 also depicts the 5G core elements which take part in a rather long chain of the key derivation. The tasks of these functional elements are summarized below.

ARPF (Authentication Credential Repository and Processing Function) stores the user's permanent K and private PKI keys, and creates the authentication vectors.

USIM (Universal Subscriber Identity Module) is the SIM (Subscriber Identity Module) application residing in the hardware-based UICC (Universal Integrated Circuit Card). It is oftentimes generalized in practice as a "SIM card". The traditional removable card is still valid in 5G. Nowadays, the UICC can also be permanently embedded into the device, or integrated into the 5G chipset. It stores securely the subscriber's permanent key K and private key of the PKI key pair, and processes cryptographical tasks together with the Mobile Equipment in order to protect the communications.

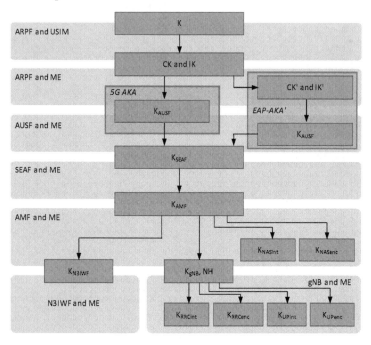

Figure 16 The 5G key hierarchy as defined by 3GPP in TS 33.501.

AUSF (Authentication Server Function) manages the authentication for the 3GPP access via 5G radio and untrusted non-3GPP access such as Wi-Fi access point.

SEAF (Security Anchor Function) is collocated within the AMF as mentioned in the 3GPP Release 15. In the primary authentication, the SEAF forms the unified anchor key K_{SEAF} which is the same for all the access scenarios. This key protects the communications of the UE and the serving network.

AMF (Access and Mobility Management Function) has many roles from which the access authentication and authorization are related to the security functions.

Table 7 The 5G keys.

Key	Description
K, CK, IK	The long-term user-specific, unique key *K* is stored securely in the USIM application of hardware-protected UICC of the user. It also is stored in ARPF, and is not moved away from it or from USIM under any circumstance. Instead, both form short-term ciphering key *CK* and integrity key *IK* for the authentication of the 5G subscriber and as a base for further key derivation.
CK', IK'	Authentication keys derived from *CK* and *IK* if EAP-AKA' are applied for Wi-Fi -based access.
K_{AUSF}	Key for AUSF in home network. AUSF generates K_{AUSF} from the authentication material (*K*, *CK* and *IK* for 5G AKA, or from CK' and IK' in case EAP-AKA' is used for, e.g., Wi-Fi access).
K_{SEAF}	Anchor key. AUSF and ME derive it from K_{AUSF}.
K_{AMF}	Key for AMF. K_{AMF} is derived from K_{SEAF}.
K_{NASint}	Integrity key for Non-Access Stratum (NAS) signaling (refers to dialogue between the mobile equipment and core network nodes, and it is access-agnostic). AMF and ME derive it from K_{AMF}. It protects NAS signaling by applying integrity algorithm.
K_{NASenc}	Encryption key for Non-Access Stratum (NAS) signaling, i.e., on access-agnostic domain. AMF and ME derive it from K_{AMF}. It protects NAS signaling by applying encryption algorithm.
K_{N3IWF}	Key for non-3GPP Inter-Working Functions. The N3IWF receives K_{N3IWF} key from the AMF, and uses it for IKEv2 between the UE and N3IWF in the procedures for untrusted non-3GPP access such as Wi-Fi.
K_{gNB}	Key for 5G gNB. AMF and ME derive it from K_{AMF}. gNB and ME obtain it for performing horizontal or vertical key derivation.
K_{RRCint}	Integrity key for Radio Resource Control (RRC) signaling. gNB and ME derive it from K_{gNB}. It protects RRC signaling with integrity algorithm.
K_{RRCenc}	Encryption key for RRC signaling. gNB and ME derive it from K_{gNB}. It protects RRC signaling with encryption algorithm.
K_{UPint}	Integrity key for User Plane (UP) data traffic. gNB and ME derive it from K_{gNB}. It protects UP traffic between ME and gNB with integrity algorithm.
K_{UPenc}	Encryption key for UP traffic. gNB and ME derive it from K_{gNB}. It protects UP traffic between ME and gNB with encryption algorithm.
NH	Intermediate key referring to Next Hop. AMF and ME derive it for providing forward security.

N3IWF (Non-3GPP Interworking functions) manages access networks (AN) which are not 3GPP-defined 5G radio access networks. Typical example is Wi-Fi.

gNB (Next generation Node B) is the 5G "base station". It receives the ciphering and integrity keys from the key derivation procedures and protects the radio interface.

In addition to the keys summarized in Table 7, there are yet more intermediate keys. K_{gNB*} is one of these; the Mobile Equipment (ME) and the 5G gNB obtain it via the Key Derivation Function (KDF). Please note that the combination of the ME and UICC and its USIM application jointly form a UE (User Equipment). The UICC/USIM handles part of the security-related tasks such as storing the K while the ME takes care of part such as the K_{gNB*} derivation.

The K'_{AMF} is yet another 5G key. Both the AMF and ME derive it via the KDF when the UE moves from one AMF to another. This case is called inter-AMF mobility scenario.

5G has adapted the widely used PKI (Public Key Infrastructure) scheme for the identifier protection based on the X.509 certification authority (CA) procedure. The PKI is a set of roles, policies and procedures to create, manage, distribute, use, store, and revoke digital certificates.

The PKI makes the public-key encryption, and it protects the 5G communications. As stated in (ETSI, 2018), the user's home network provisions the public key. It is securely stored in the user's USIM whereas the ARPF houses the respective private key.

The subscriber's privacy, provisioning and updates of the public key are controlled by the home mobile network operator.

Further reading	
▶3GPP	*TS 33.501* (5G security architecture and procedures)
▶5G Explained	Section 8.9.2 (key hierarchy)

Location Based Services

Positioning will play an elemental role in the 5G networks. 5G enables a variety of location-based services and applications such as intelligent traffic system (ITS) and self-driving cars.

5G supports an integrated Location Based Service (LBS) which provides tools for many new use cases. (Tampere University of Technology, n.d.) Figure 17 depicts the 5G positioning architecture for the NG-RAN and E-UTRAN as described in the 3GPP TS 38.305. The main element for the 5G LBS is the LMF (Location Management Function). The radio access network may include a set of evolved 4G ng-eNBs, 5G gNBs, or both.

TP is a *Transmission Point* which refers to a set of geographically co-located transmit antennas for a single cell, a part of a single cell, or a single PRS-only (Positioning Reference Signal for radio network). The Transmission Points can be formed by antennas of ng-eNB or gNB, remote radio heads, remote antenna of a base station, or antenna of a PRS-only TP. In practice, each transmission point may correspond to a cell.

Either the GMLC (Gateway Mobile Location Center) or the target UE itself may trigger a location-based request. The network routes the request to the AMF (detailed in Core Chapter).

Also, the AMF is able to trigger the LBS request for a target UE in special situations such as when a user dials an IMS (IP Multimedia Subsystem) emergency call.

The AMF sends a location service request, accompanied by any optional assistance data, to the LMF which processes the data. The LMF sends the resolved UE's location back to the AMF.

As depicted in Figure 17, the LMF may signal with an E-SMLC (Evolved Serving Mobile Location Centre), SLP (SUPL Location Platform), or with both.

E-SMLC	SLP
The E-SMLC provides the LMF with an access to an OTDOA (Observed Time Difference of Arrival) information. The OTDOA reasons the user's location based on the signaling from a set of base stations. The more signals from different base stations the UE receives, the more accurate the positioning.	The SLP is based on SUPL (Secure User Plane Location) which provides user's positioning over the data link. The SLP contains an SLC (SUPL Location Center) and SPC (SUPL Positioning Center) which calculate the UE's position from satellites and their positioning database, Wi-Fi and/or cellular networks.

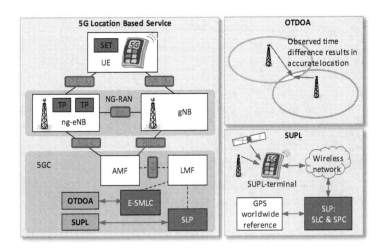

Figure 17 5G positioning architecture.

Further reading	
▶3GPP	*TS 38.305* (UE positioning); *TS 29.516* (E-SMLC)
▶ETSI	*TS 129 572* (Location management systems)
▶5G Explained	Section 7.2.8 (Location-Based Services)

Massive Internet of Things

Massive Internet of Things, or mIoT, also known as massive Machine Type Communications (mMTC), refers to the ability of 5G to deliver contents among a vastly increased number of devices communicating simultaneously such as intelligent sensors.

5G has been designed taking into account the forthcoming volume of the mIoT. In fact, the ITU-R IMT-2020 requirement set includes a value for the number of simultaneously communicating mIoT devices, which is 1 Million devices per km^2.

The generic term for the IoT communication system is LPWAN (Low Power Wide Area Network). The name indicates that the respective terminals are highly optimized to support long battery lifetime. Oftentimes, the communication of such devices is based on low bit rate and occasional transmissions.

The communications for Internet of Things can be referred to as C-IoT (Cellular IoT) when mobile communication network is involved. There are also alternative technologies developed such as LoRa, SigFox and RPMA. The factor that distinguishes the C-IoT is the presence of IMEI (International Mobile Equipment Identity) which is not found in the alternative devices for their own system.

The evolution of C-IoT, as defined by 3GPP, has been already going on since the LTE Release 8. There are few variants in the markets based on the LTE as summarized in Table 8. There also is a mode defined for GSM, EC-GSM-IoT (Extended Coverage GSM IoT) which is defined in Release 8.

Table 8 C-IoT categories for LTE.

Technical data	Cat 1 LTE Rel. 8	Cat 0 LTE Rel. 12	Cat M1 LTE Rel. 13	NB-IoT LTE Rel. 13
DL peak data Mb/s	10	1	1	0.2
UL peak data Mb/s	5	1	1	0.2
Complexity	High	Medium	Low/Medium	Very low

LTE-M (or Cat M1) is an LTE machine-type communications (MTC) LPWA standard as per 3GPP Release 13. LTE-M device complexity is low, and it supports massive connection density, low device power consumption, low latency and extended coverage allowing the reuse of the LTE installed base.

NB-IoT (Narrowband IoT) is a 3GPP Release 13 solution that provides improved indoor coverage, support for massive number of low throughput devices, low delay sensitivity, low cost, and low device power consumption.

As stated in (GSMA, 2018), NB-IoT and LTE-M support both the LTE and 5G LPWA requirements. 3GPP has agreed that the LPWA use cases will continue to be addressed by evolving NB-IoT and LTE-M as part of the 5G specifications. The NB-IoT and LTE-M defined prior to the 5G era are thus considered to be part of the 5G family; in fact, one of the 5G NR deployment scenarios is to place the LTE-M or NB-IoT directly into a 5G NR frequency band.

Further reading	
▶3GPP	*TR 36.802* (NB-IOT radio transmission and reception)
▶5G Explained	Section 3.3 (the role of 3GPP in LPWA and IoT)

Measurements

In addition to the measurements during the setup of the 5G network, one of the essential tasks of mobile network operator is to measure periodically the network performance in the operational phase. The results form an important base for constant optimization of the network.

The measurement equipment for the previous generation networks serve still largely for the 5G, too. Nevertheless, as 5G includes many new functionalities, new hardware is needed to support them. It might be possible to upgrade old equipment if the functionality is software-based, which is the case for, e.g., protocol analyzer with upgraded 5G protocol stacks. In some other cases, if more processing power or extended frequency ranges are required, hardware upgrade is necessary to support wider bandwidths and increased data speeds.

Operators typically perform regular coverage area measurements to ensure the planned radio performance is achieved, and to monitor any potential issues. 5G is based on the already existing frequency bands as well as completely new, remarkably higher frequencies in the mm-Wave band.

The principles of the radio measurements remain the same, and many of the old measurement types are still relevant in 5G such as quality (bit rates, bit error rates), radio coverage areas (received power level) and capacity. Nevertheless, 5G increases significantly radio bandwidths which new measurement equipment need to support. The 5G bands will be supporting values typically from 500 MHz up to 2 GHz, and the novelty 5G bands are between 6 and 60 GHz.

Table 9 Examples of 5G measurement equipment.

Type	Examples and suitability
Signal analyzer	Signal and spectral analyzer are designed to support frequency ranges of several tens of GHz, and in many cases, the range can be further extended by down-conversion. The demodulation of bandwidth may be several hundreds of GHz in the basic models while the range can be extended up to the maximum of 2 GHz for signal analysis by native HW/SW or by relying on, e.g., external oscilloscope. The equipment may analyze digitally modulated singe subcarriers in bit-level or limited to generic OFDM signal analysis. The supported frequency bands may vary from basic models' sub-6 GHz up to highest foreseen bands in the 100 GHz range.
Signal generator	The vector signal generators serve in simulation of wideband signals and they may have native 2 GHz modulation bandwidth. The equipment may also be able to generate a variety of wave forms, apart from the 5G OFDM, and generation of inter-effects and multiple component carriers for carrier-aggregated simulations. Modulation schemes can include also 256-QAM. Channel power level, co-channel power and error vector magnitude may be included, too.
OTA	OTA (Over-the-Air) test chamber can be fixed or mobile for passive and active 5G antenna measurements, including RF patterns of complex antennas in 3D domain. The system may also support radio transceiver performance measurements. Typically, these types of measurements support all the defined radio frequencies up to the maximum, around 87 GHz. The measurements indicate the impact of 5G RF OTA parameter values.
Network emulator	The network emulator can be useful for the performance measurements of multiple subchannels of the OFDM, combined with vector signal generator, as well as signal and spectrum analyzer.

For the core and laboratory tests, wideband *signal generators* and *signal analyzers* are useful tools. Other measurement types relate to modulation, energy efficiency indicated by Peak-to-Average Power Ratio (PAPR), spectral efficiency indicated by the Out of Band leakage (OOB), reliability of the link indicated by Bit Error Rate (BER), MIMO performance, transmission latency, and quality of the synchronization, to mention some. Table 9 summarizes some of the key measurement types in 5G.

5G era is approaching fast as the Non-Standalone 5G deployments are currently taking place in many countries. The next phase is the Standalone (SA) option, which will provide with the native, full 5G user experience.

Nevertheless, the schedule for the fully and globally interoperable IMT-2020 system, and its candidate technologies as evaluated by ITU-R, is still the original 2020 as the term indicates. IMT-2020 is a set of highly demanding requirements for the next generation systems that will be forming the base for a connected society, and the 5G specifications as defined by 3GPP will definitely form part of this environment. It is worth remembering that the new frequency bands for 5G will be discussed and decided at the ITU-R World Radio Conference in 2019 so we'll need to wait still some time until the complete 5G system standards are ready.

Meanwhile, there are practical proof of concepts for 5G Release 16, as well as respective measurement equipment to make all the trials and pilots possible. The following presents a snapshot of the recent activities in the field. (Penttinen J. , 2017)

Anritsu has developed their Universal Wireless test platform to support new sub-6 GHz bands aimed for the research of the New Radio (NR) of 5G.

Keysight has extended the PathWave software platform for 5G simulations, design and test workflows. It works as a base for new innovation and product development of the customers. In addition, Keysight has developed low-frequency noise analyzer especially suitable for the IoT sensor measurements.

National Instruments (NI) has productized PXIe-5840 radio tester that is suitable for the Non-Standalone NR measurements up to 6 GHz frequency bands. The company has also cooperated with Samsung on 28 GHz 5G equipment and performance tests.

Rohde&Schwartz (R&S) has developed a movable OTA test chamber for 5G systems, covering the test cases for antenna and transceiver measurements. R&S has also developed NR measurement equipment supporting wide bandwidths, network emulators, vector signal generators and other relevant test devices for 5G.

Siemens has extended knowledge on 5G-related simulations, offering services for base station and silicon vendors for testing 5G functionality and performance. This has been possible by acquiring specialized companies such as Sarokal.

Testing of Release 16 performance is essential to expedite deployment of new 5G equipment as the second phase standards are approaching the final stage. There are still many challenges such as the accuracy of the current measurements basing on partially simulated environments but these can be coped with by adjusting the final products accordingly when the standards are ready.

Further reading	
▶3GPP	*TS 34.114* (User Equipment / Mobile Station Over the Air antenna performance; conformance testing)
▶5G Explained	Section 10.9.2 (measurement principles of 5G)

Network Functions Virtualization

5G relies on Network Functions Virtualization (NFV). It refers to the decoupling of software from hardware, and it provides 5G mobile network operators (MNO) with the possibility to deploy service-based architecture model.

The service-based architecture facilitates highly modular and reusable approach for the 5G network deployments. It allows MNOs to deploy the 5G gradually. The service-based architecture contains the core network functions by interconnecting them with the rest of the system as depicted in Figure 8 of Core Network Chapter. NFV also enables distributed cloud for optimizing further the deployment of the services in a flexible way. Distributed cloud refers to the possibility to treat a variety of data centers as a single, virtual data center.

NFV makes it possible to replace mobile network functions on dedicated elements such as Equipment Identity Register (EIR), as well as other appliances such as routers and firewalls, with virtualized instances in a form of a software running on COTS (commercial off-the-shelf) hardware.

The network functions virtualization simplifies the operation of the functions by decoupling them from the traditionally required stand-alone hardware components. The virtual network functions are deployed on high volume servers or cloud infrastructure instead of specialized hardware. This model expedites the cost-efficient commercial deployments.

NFV also enables the 5G Network Slicing providing MNOs with the possibility to customize and optimize resources for different type of verticals. In addition to the network slicing as such, 5G NFV makes it possible to divide physical network into a set of virtual networks which may support multiple radio access networks.

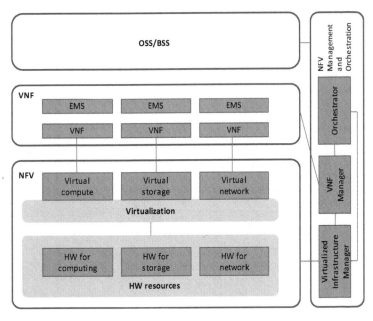

Figure 18 The principle of NFV.

In this concept, the underlying Network Functions Virtualization (NFV) acts as a layer between hardware resources and virtual environment such as virtual compute, storage and network. The NFV is connected to the Virtual Network Functions (VNF) which are controlled by Element Management System (EMS). Furthermore, the NFV management is connected to the MNO's operations support system / business support system (OSS / BSS) which, among other tasks, takes care of the 5G network maintenance and billing.

The benefits of the Network Functions Virtualization include also the possibility to optimize further resource provisioning of the virtual network functions basing on the cost or other criteria and scale the virtual network functions accordingly. 5G MNO can use NFV to introduce new services as per need.

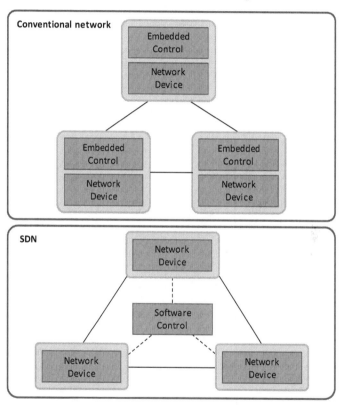

Figure 19 The principle of SDN and comparison with traditional network.

Software Defined Networking (SDN) is a network architecture model to overcome with hardware limitations. SDN is related to NFV as it benefits in the abstraction of lower level functions, and instead moves them to a normalized control plane which manages network via the APIs (Application Programming Interface) as depicted in Figure 18.

SDN thus provides the MNOs with a possibility to offer 5G services via a centralized control plane in a hardware-agnostic manner. It enhances the data flows via lower bandwidth and lower latency, and also the network redundancy can be managed more efficiently.

SDN results in more flexible networks. The SDN network architecture supports the 5G ecosystem requirements and it can be used to design, build and manage 5G networks. As the control and user planes are separated, the control plane is directly programmable while the underlying infrastructure is abstracted for applications and network services, to create various network hierarchies.

While the control is distributed in the traditional networks, SDN detaches the control plane from the network hardware, enabling packet data flow control through a controller. The controller is located between network devices and applications as depicted in Figure 19. As a consequence, network control is programmable, facilitating the management of the 5G network as well as modification and addition of services. (SDX Central, 2019)

Further reading	
▶ETSI	*Network Function Virtualization Industry Specification Group*
▶5G Explained	Section 4.3.2 (Network Function Virtualization)

Network Slicing

5G uses network slicing which refers to a "networks within network" concept. Network slicing is a technology that allows operator to form a number of logical networks on top of a common physical infrastructure to personalize service levels.

Network slicing is a new technology which allows 5G operators to use their physical mobile network partitioned into multiple virtual networks. Different customer segments, or verticals, benefit from this solution because they are able to receive focused support for the use of different type of services. This is one of the key differentiators of 5G compared to the legacy systems. By deploying network slicing feature, the operator may configure each slice for better complying with the requirements of different user segments.

Operators may use network slicing for supporting an as-a-Service model for the customers. It enhances operational efficiency and speeds up time-to-market for new services.

Network slicing technology is possible thanks to the software-defined networking, network functions virtualization and advanced 5G network orchestration. In fact, these features are essential to apply and manage the network slicing.

The operator can form, manage and terminate network slices upon need in a highly dynamic fashion, and each slice can be optimized for a specific use case. This is a big difference to any previous systems in which the mobile network is adjusted in a more static manner and offers certain uniform performance to be shared amongst the users in that area.

Instead, 5G network slices can be tuned to offer different user experiences within a selected area. Figure 20 depicts the principle of the network slicing, each using only the needed Network Functions (NF) per each slice.

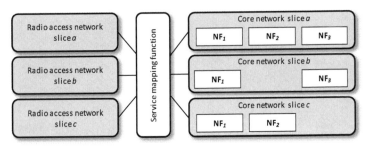

Figure 20 The principle of the Network Slicing in 5G.

Furthermore, the 5G user equipment is capable of joining one or multiple network slices at the same time supporting different parallel services.

As an example, smart city sensor might not require the highest data speeds but can benefit from the possibility to share low bit-rate resources amongst a big number of other devices communicating simultaneously, so the sensors can subscribe to a massive Machine Type Communications (mMTC) network slice offering such desired performance. Meanwhile, some other users within the very same area, such as critical communications applications, may want the highest reliability, so operator can offer an adjusted Ultra-Reliable Low Latency Communication URLLC network slice for them. Figure 21 depicts this principle.

The benefit of this approach is that differently adjusted network slices optimize the network resources while providing a good user experience.

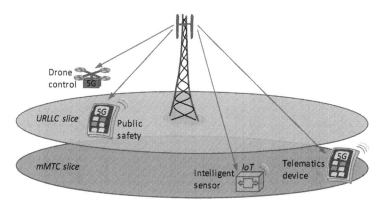

Figure 21 Network slice forms virtual networks optimized for different users.

The network slicing offers thus important benefits such as dynamic and customizable end-to-end operations to fulfill special needs of highly versatile users – including traditional subscribers and new verticals alike – such as law enforcement entities, drones, VR/AR industry and autonomous vehicles.

Being highly dynamic, network slicing requires advanced integrated and automated techniques to handle the flexibility and efficiency. The development of the supporting solutions includes areas such as artificial intelligence which eases the service orchestration throughout the life cycle of the slices, from concept testing, commercial operations and maintenance up to the end of the product.

Each 5G network slice equals to a separate, self-contained virtual network for specific use cases. Each slice can have thus separate Service Level Assurance (SLA) determined individually per each customer with the service provider. This is the reason why the service management and service assurance need to be divided into each network slice separately.

The lifetime of some of the slices may be quite short in practice, such as in some special, temporal events. The challenge of it is that the same resource layer is used as a basis for multiple slices, and it might not be possible to comply with the original assurance of adequate resources per slice due to unforeseen peak capacity demand during the event.

The solution for this issue is the deployment of a service assurance tool that can correlate service experience with the respective resource use. In an optimal setup, this guarantees that a single slice does not consume more resources than is absolutely needed to fulfil the respective SLA, considering the performance and capacity.

The strategy for such balancing depends on the use case, so the service priority for a set of non-critical mIoT sensors is lower than for applications requiring ultra-low latency. This type of balancing is known as contextual service assurance between the network management and orchestration (NMO) plane and the slice layer.

For the most efficient network slicing benefiting the ecosystem, including the national and international roaming between mobile network operators, it is important to consider common principles that ensure fluent user experiences in a global scale. GSMA Future Networks is working on the forming of a generic network slice template (GST) which presents attributes for key services relying on network slicing. These attributes refer to the technical parameters providing desired performance level for the slices. The document also shows typical examples of respective Network Slice Type (NEST) database and proposes recommended minimum set of attributes and their feasible values to be applied by mobile network operators.

Further reading	
▶GSMA	*NG.116*, Generic Network Slice Template (GST)
▶5G Explained	Section 4.3.1 (Network Slicing)

Open Source

The 5G architecture uses the principles of the Open Source. Open Source is something people are able to share, modify and use based on available design for anyone. As for software, the Open Source refers to a possibility to share it, investigate the code and its functions, modify, copy and distribute it upon the SW licensing terms. This principle helps to expedite the 5G development.

The Open Source Initiative (OSI) license is an example of the Open Source code environment. Of the 5G operators, AT&T decided to implement an open source platform based on cloud environment of Kubernetes and OpenStack. (Wagner, 2019)

The 5G RAN disaggregates 4G eNB and 5G gNB functionalities into modules such as Distributed Unit (DU), and Centralized Unit separated into User and Control Planes (CU-CP and CU-UP), and defines standardized, interoperable interfaces. The aim of this work is to optimize further the radio resource use and load balancing of separate, HW-agnostic signaling and user data processing in virtualized environment.

O-RAN has been involved with the work in cooperation with 3GPP. (O-RAN Alliance, n.d.) The goals of this effort is to leverage open source implementations, and to speed up the development and deployments.

Other areas, in form of workgroups of O-RAN, include the following activities:

- Non-real-time RAN intelligent controller and *A1* interface for radio resource management, procedure and policy optimization, and Artificial Intelligence (AI) and Machine Learning (ML) models;

- Near-real-time *RIC* and *E2* interface architecture on decoupled SW implementation of control plane;

- Stack reference design and *E1*, *F1* and *V1* interfaces, multi-vendor profile specifications for *F1*, *W1*, *E1*, *X2*, and *Xn* interfaces;

- Open Fronthaul Interfaces for promoting multi-vendor DU-RRU (Distributed Unit – Radio Remote Unit) interoperability;

- Cloudification and orchestration for decoupling RAN software from the underlying hardware platforms;

- White-box hardware to reduce cost of 5G deployment via reference design of decoupled software and hardware.

Examples of some other stakeholders and systems promoting Open Source are: Open Compute Project (OCP) which provides telecom data center operators with open platform standards, Disaggregated Network Operating System (DANOS) which is open networking operating system, and P4 which is open-source initiative for interacting with networking forwarding planes. (5G Americas, 2019)

Further reading	
▶O-RAN Alliance	*O-RAN Alliance Web page* (O-RAN Alliance, n.d.)
▶5G Explained	Section 4.3.3 (Open Source)

Patents and IPR

As 5G is becoming reality along with the deployment of the initial networks throughout 2019, it opens room for many new business opportunities, too. The commercial side of 5G include also patents which help many companies to monetize the efforts of the research and development.

One might wonder, who makes money with 5G patents? As relevant question as it is, the answer is not completely clear quite yet. The reason is that – as the 5G services are merely starting to take off – many of the patents are still pending in various countries. This evaluation phase is rather confidential, so the exact figures on the proportion of stakeholder's essential patent portfolio can only be guessed at the moment.

Nevertheless, as indicated in *"Who is leading the 5G patent race? A patent landscape analysis on declared SEPs and standards contributions"* by IPlytics (IPlytics, 2019) Chinese companies are estimated to have 34 % proportion of the 5G-related patents compared to other stakeholders.

According to the same source, 5G will be an elemental component for the IoT connectivity of many objects and vehicles. Thus, related sectors relying on 5G, such as transport, energy, manufacturing, healthcare, and entertainment, are using the patented solutions and contribute to the 5G business via the licensing fees.

In order to understand the impacts in early stage, standardization communities require IPR (Intelligence Property Rights) declarations from the organizational partners. As an example, the 3GPP Project Coordination Group (PCG) is responsible for maintaining a register of IPR declarations relevant to 3GPP.

According to the 3GPP rules, individual members declare to their organizational partners IPRs which they believe are essential, or potentially essential, to any work being conducted within 3GPP. Furthermore, a call for such IPRs is done during 3GPP work group meetings that lists IPR declarations sorted by Organizational Partners. (3GPP, 2019)

As soon as the patents become public, their impacts can be understood better. The patent holders include network element and device manufacturers, mobile network and service operators, chipset and module manufacturers, research institutes and academia, and other companies and organizations who might have contributed to the 5G evolution protecting their IPRs. Nevertheless, as the new patent evaluation may take months or up to several years, we have to still wait for the final answers.

Then why is this topic relevant in the first place? The simple reason is that patents generate more business opportunities to their owners as they provide concrete means to monetize the innovations. The *essential patents* are in a key position as they describe the critical functionalities of 5G. In fact, without the adaptation of solutions described in the essential patents, 5G might not work as standardized. Furthermore, as stated in (IPlytics, 2019), the patents can be highly lucrative business.

The ones owning essential patents can license their solutions while other stakeholders pay for the right to deploy and use them. Many companies do have essential patents so there is an established business model for the cross-licensing of the patent portfolios balancing the gains and costs of IPR-protected solutions. The ones owning more patents than others will gain in this business while the ones with lighter or non-existing patent portfolio are the net payers.

How profitable the licensing is, then? That depends on the agreements between business partners. The topic is business-confidential by default, but as an example on smart devices, the proportion of the licensing costs may represent 20-25% of the total price of the device according to Ref. (Ann Armstrong, 2014).

The IPRs can be important for the commercial success of the companies, and may even have impact on their stock market prices. It is thus easy to guess that patent portfolios are especially interesting information for the investors.

It might be possible to estimate patent portfolios researching the overall activity of different companies at the 5G standardization bodies because the actively contributing companies may have invented and protected respective new ideas. The contribution happens in working groups of 3GPP and other relevant standard developing organizations and industry forums. The information on the contributions, including the change requests and new item proposals, is public and can be investigated from the web portals such as *www.3gpp.org*.

The patent portfolio has been an increasingly important business item already for long time during the existence of the previous mobile generations. Every now and then, there have been legal disputes as companies want to protect their IPRs. It is not hard to guess that the role of patents will be at least equally important also in 5G era.

For us users, the IPR business can probably be seen most concretely in increased sales prices of mobile devices. Also, other stake holders such as mobile network operators may need to compensate their indirect IPR cost impacts with subscription or other service fees.

Regardless of the price impact, the IPR business is an elemental part of the industry that keeps companies innovating interesting devices and services in a fast pace which is, after all, beneficial for us all.

Further reading	
▶3GPP	*3GPP Legal Matters* (https://www.3gpp.org/about-3gpp/legal-matters)

Planning of 5G Network

The radio network planning balances the capacity, coverage and quality. The core and transport networks, in turn, need to support the generated traffic from the radio network to avoid bottlenecks while the commercial aim is to optimize the costs.

The **radio network dimensioning** is based on the radio link budget and radio wave propagation prediction models. They help operators to estimate the maximum feasible distance from the base station up to where the user equipment can still be served in different topological environments.

The radio waves attenuate more in dense urban areas due to obstacles such as high buildings. The coverage is largest in open areas as the attenuation is lowest for the Line of Sight (LOS) scenarios. Also, the higher frequencies attenuate more than the lower ones. The lower bands are thus used typically in large rural and sub-urban areas, but on the other hand, the vegetation and trees may decrease their radio propagation range. The radio network planning is thus a complex task which requires theoretical and practical knowledge about radio waves, field measurements, tools and simulators.

The 5G radio wave propagation prediction models are based on already existing and new methods. For the frequencies above 6 GHz, which is a new area compared to the previous generations, the impact of obstructing materials is rather significant. The 60 GHz band is a special one as the oxygen molecules cause an additional 20 dB attenuation peak to the radio wave propagation which reduces further the coverage.

As explained in Frequencies Chapter, the low-band refers to sub-1 GHz, mid-band to 1-6 GHz, and high-band to frequencies above 6 GHz. In the initial phase of 5G, the high-bands on millimeter-wave area of 24 GHz, 28 GHz and 38 GHz are assumed to be popular in the USA whereas Europe and China use 26 GHz. There are also many frequency range candidates on lower bands such as 600–900 MHz, 1.5 GHz, 2.1 GHz, 2.3 GHz and 2.6 GHz to be evaluated for the 5G deployments. These bands are especially useful in applications requiring less capacity such as IoT communications as they provide a rather large coverage.

Thanks to the radio wave propagation characteristics, the mid-bands of 3.30–4.20 GHz and 4.40–4.99 GHz are especially suitable for the initial phase of the 5G. They provide a feasible compromise for the capacity and coverage, and facilitate the expedited 5G deployment schedules.

The same old principles of the radio frequency (RF) propagation applies to 5G; the higher the frequency is, the smaller the coverage area. The basic rule suggests that when a frequency doubles, the received power level lowers by 3 decibels – which equals to a 50 % reduction of the original power. Thus, thanks to their wider bandwidth, the frequencies above 6 GHz are mostly adequate for delivering high radio capacity in small cell environments. These cells are oftentimes limited to very short distances in outdoors, or within single floors in indoors, which makes them especially suitable for a dense city environment.

Table 10 Examples of typical 5G coverage areas.

Frequency Band	Typical outdoor ranges (can vary largely in practice)
Below 1 GHz	Few miles up to ten miles
1 – 6 GHz	One mile up to few miles
Above 6 GHz	Few hundred feet up to one mile

Along with the increased complexity of 5G and high dependency on the traffic types, geographical topology, and other factors, its radio link budget is typically based on radio network planning simulators and digital cluster maps. Nevertheless, by applying adequate propagation models, it is possible to make a rough estimate on the expected cell sizes with only basic tools. This type of exercise may be feasible in the nominal planning phase to estimate the rough number of the base stations in the planned area, and the respective high-level investment.

The radio link budget takes into account the very key aspects in the communication link, such as transmitter power, antenna cable loss, and antenna gain for each beam (which can be highly dynamic in 5G). In the receiving side, the sensitivity and noise figure are some of the parameters. It also takes into account fading variations and indoor attenuation in radio reception.

The ultimate goal of the radio link budget is to dimension the expected cell size (radius) so that the planned services, such as voice calls and data, comply with the designed quality criteria. This happens by balancing both uplink (the transmission from user device to base station) and downlink (the opposite direction) so that the desired services, such as 2-way voice call, can be used adequately.

As depicted in Figure 22, the closer the user is located to the base station, the higher the 5G data speed. The higher received power helps to lower the data error rate. The final bit rate per device depends on the modulation scheme and the number of the devices communicating simultaneously in the cell area. To prioritize and balance between different services and users, 5G has an advanced Quality of Service classification.

The correct dimensioning of the radio link budget is of utmost importance for the mobile network operators as it indicates how dense the network will be – which, in turn, dictates the initial capital expenses (CAPEX) and longer-term operating expenses (OPEX) of the network. A well optimized radio network can ensure adequate return of investment (RoI) and sustainable business while poorly planned radio network may waste energy, money, and capacity.

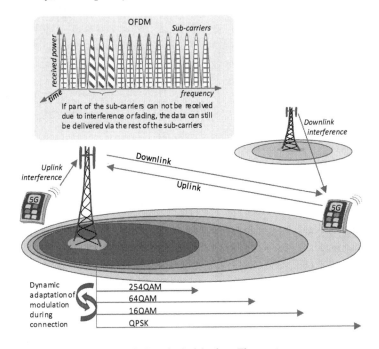

Figure 22 The principle of 5G radio link budget. The maximum coverage area depends on the modulation scheme, among other factors. The 254QAM provides 5G users with the lowest coverage but the highest data speeds.

As an example, it is important to balance the power levels of the useful and interfering transmissions instead of increasing the power levels in every site. A skillful radio engineer who knows by heart the theories and practices of the radio propagation, network planning, optimization and measurement techniques, is one of the most invaluable assets of mobile network operator or cooperating party planning the 5G networks.

114

Optimized radio network may save lots of money by reducing unnecessary transmitter power as it decreases interference. Figure 22 presents the principle of the interference sources in 5G. The interference may be caused by nearby user equipment to the base stations, or the nearby base station can generate interference to the user equipment. The 5G OFDM (Orthogonal Frequency Division Multiplexing) is an adaptive multi-subcarrier technique which tolerates well such interferences, adjusting accordingly the data throughput.

The radio communication links are highly dynamic. The radio network must be optimized constantly based on subscription forecasts, field measurements, and automatized network analytics. As 5G may be based on a significant number of small cells operating in mm-Wave bands, especially in dense city areas, one option is to deploy radio elements on light poles which are connected to the core network via fiber optics or radio links.

For the core network planning, the key task is the correct dimensioning of the capacity and reliability of the infrastructure, taking into account the future outlook for the needed traffic.

In any network planning, the core network must ensure adequate capacity and performance for the customers. The quality can be measured by the service uptime which typically requires redundancy to minimize any single point of failures. In the strictest environments, geographical "active-active" redundancy may be used for the critical network elements. That is rather expensive as the elements execute the same tasks in a parallel fashion; if one fails, the redundant element takes over the tasks in real time.

Also, the timing of the investment is important in order to optimize the costs of materials, storage, and deployment efforts. A good near-term and longer-term forecast for the network utilization is thus of utmost importance.

5G core network functionality is predominantly virtualized, and it is based on a service-based architecture. Data centers will be an important part of the 5G core network. Their deployment plan considers the interconnectivity with the 5G infrastructure.

The data center operations can be outsourced, too, which will bring along with new considerations for mobile network operators, such as requirements for data center certification models, and a Service Level Assurance (SLA) for the expected quality using appropriate redundancy and recovery classes.

Transport network connects the radio and core networks. The dimensioning of the transport network becomes increasingly important because the 5G radio interface will be capable of generating considerably more traffic than was the case in the previous generations.

The transmission network can be based on fixed and wireless solutions, the fiber optics and radio links being typical technologies. For the radio links, adequate frequency planning maximizes the capacity and minimizes the interferences. The transport network may rely on increased intelligence, flexibility and automation to better cope with the high dynamics of 5G.

For the fiber optics deployment, the planning must be done well before the expected traffic reaches a critical limit. The overall 5G planning is largely related to the new cloud environment, and edge computing plays a key role in that ecosystem as described in Core Chapter.

Further reading	
▶**3GPP**	*TS 43.030* (radio network planning aspects); *TR 38.901* (channel modeling study)
▶**ETSI**	*TR 138 913* (5G radio link budget and scenarios)
▶**5G Explained**	Chapter 9 (5G network planning and optimization)

Quality of Service

The integrated Quality of Service (QoS) mechanism of 5G is an enhanced version from the previous generations. 5G customers benefit from this evolution as the services can be used in a more flexible and efficient manner.

For the fluent user experiences, it is essential that customers have seamless IP session continuity when the device is moving within the service area. In the LTE system, this refers to the maintaining of the IP address of PDN Gateway (P-GW) and Packet Data Unit (PDU) session while on the move. It should be noted that there are also applications which do not require the actual IP session continuity for the smooth experience.

The 5G QoS is based on further developed QoS flows for guaranteed bit rates (GBR) and on the flows not requiring guaranteed bit rates. This approach provides QoS differentiation for PDU sessions at signaling plane level.

5G provides a set of different types of session continuity modes coping well with the UE and service types. 5G QoS is designed to differentiate data services, and to adapt the performance requirements of a variety of applications. This is important because – regardless of the considerably increased capacity of 5G – the radio resources are limited as soon as users demand data from the network simultaneously.

The 5G QoS supports different requirements of other access technologies, too. Wi-Fi access points are an example of these.

5G can cooperate with them thanks to the integrated, native 5G security functions for such scenarios.

Operators can deploy application functions in a flexible manner. 5G deploys the QoS via a set of of Session and Service Continuity modes (SSC). The SSC modes provide applications with a possibility to influence in the selection of adequate data service characteristics.

As depicted in Figure 23, there are 3 modes in 5G.

SSC 1 is a familiar mode from the LTE networks. It ensures that the IP anchor remains stable supporting applications and maintaining the link for UE in location updates.

SSC 2 is a new, "break-before-make" mode. It means that a network may break the connectivity and release the data session and possibly the IP address before creating a new connection.

SSC 3 is another new model, which can be characterized as "make-before-break". It means that the network makes sure that the 5G device does not lose the connectivity. In this mode, before breaking the previous connection, the network creates the new connection thus allowing the service continuity. The IP address will change in this mode.

Both SSC 2 and SSC 3 facilitate the relocation of the IP anchor in the 5G environment.

Once the SSC mode is associated with a PDU Session of 5G it would not change during the rest of the lifetime of the PDU Session. The 5G architecture allows thus applications to influence the selection of SSC modes as needed for required data service.

5G also has QoS-solutions called *Uplink Classifier* (UL-CL) and *Branching Point*. The 5G UPF supports UL-CL and it is designed to divert traffic to local data networks based on traffic matching filters of UE traffic.

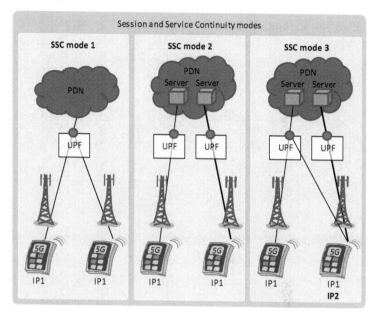

Figure 23 Principle of 5G SSC modes.

The Branching point, in turn, refers to UPF's generalized logical data plane function for the UE PDU session.

Both UL-CL and Branching Point methods allow the injection of traffic selectively to and from application functions on the user plane links.

The QoS maps with the users Quality of Experience (QoE) although their definitions differ. While the QoS refers to the technical performance of the system and connectivity, the QoE is more customer-focused measure and provides understanding on the human interpretation of the quality of the service. It indicates the customer's "happiness" – or annoyance – based on subjective experiences related to a service such as voice call or Internet browsing.

QoE thus indicates the quality level of the whole service experience. The relation of QoS and QoE can be seen from the fact that a poor QoS lowers also the perceived QoE.

Further reading	
▶3GPP	*TS 38.300* (New Radio); *TS 23.501* (system architecture)
▶5G Explained	Chapter 4.4.3 (QoS in 5G)

Radio Network

5G brings along with many new frequency bands, including mm-Wave spectrum which brings more capacity for the users. 5G has also many enhanced and new radio-related solutions.

The capacity of the radio interface can be increased widening the radio frequency (RF) bandwidth. As the low-band and mid-band frequency ranges are already rather occupied, the high-bands will bring the highly needed additional capacity for the 5G users thanks to the considerably wider bandwidths of the mm-Wave regions as can be seen in Figure 13 of Frequencies Chapter. The downside of the higher frequency bands is the more limited radio propagation. The coverage area of the mm-Wave bands might be up to some hundreds of feet while the lowest bands are adequate for large, sub-urban and rural areas – with the cost of less data throughput, though.

5G is capable of selecting automatically the most adequate modulation scheme from the Quadrature Phase Shift Keying (QPSK), Quadrature Amplitude Modulation of 16QAM, 64QAM and 254QAM. The QPSK is the most robust mode which provides largest coverage but lowest data speeds. 256QAM is the most sensitive for interferences and works thus only in very limited areas where it provides the highest bit rates.

Both uplink and downlink of 5G are based on the OFDM (Orthogonal Frequency Division Multiplexing) whereas LTE uses OFDM in downlink and SC-TDMA (Single Carrier Time Division Multiple Access) in uplink. The 5G uplink has better PAPR (Peak-to-Average Power Ratio) performance than LTE benefiting especially low-power IoT devices.

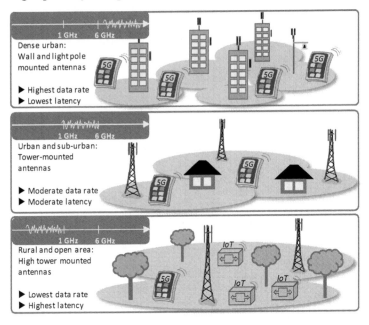

Figure 24 Principle of the high, mid and low frequency bands of 5G.

MIMO (Multiple In, Multiple Out) antenna systems are an integral part of 5G era. They provide more capacity thanks to the multiple radio paths. There will also be adaptive beam forming antennas, so instead of static cellular coverage, 5G will provide "beam coverage", a set of individual beams sweeping the area.

The MIMO concept is related closely to the transmitter and receiver antenna beamforming to improve further the performance and to limit interferences. Not only the beamforming is useful for high frequencies, but it can be assumed to form an important base for many low-frequency scenarios to extend coverage and to provide higher data speeds.

The general term for the 5G radio access network base station is NG-RAN node. It can be either a gNB (which refers to the native 5G base station) or an ng-eNB (which is the intermediate radio base station basing on the evolved 4G era). These gNB and ng-eNB elements are interconnected by the *Xn* interface within the radio network. Both elements are connected furthermore via the *NG* interfaces to the 5G Core Network. This interface is divided into two parts: User and Control interfaces.

The control interface is referred to as *NG-C*, and it connects the radio base stations to the AMF (Access and Mobility Management Function). The user interface is referred to as *NG-U* and it connects the radio base stations to the UPF (User Plane Function). These interfaces are described in the 3GPP TS 23.501 whereas the functional interface for the control and user plan split as well as the 5G architecture are explained in the 3GPP TS 38.401.

In 5G, the further optimization of the radio resources takes place by separating user and control communications. It refers to the decoupling user data and control planes. This provides means to separate also the scaling of user plane capacity and control functionality. One example of this is the delivery of the user data via a dense access node layer while the system information messages are delivered via overlaying macro layer. So, 5G gives possibility to optimize the capacity via these different paths for signaling and data by applying CUPS (Control and User Plane Separation).

This separation applies also over over multiple frequency bands and radio access technologies. In 5G, it provides possibility to deliver the user data via dense, high-capacity 5G layer on higher frequency whereas overlaying LTE system provides reliable signaling for call control. This possibility to enable mobile devices to connect both to LTE and 5G New Radio simultaneously is referred to as Dual Connectivity **EN-DC** (E-UTRAN New Radio – Dual Connectivity).

The physical layer channels of 5G overview is presented in 3GPP TS 38.201 (general description of NR) and TS 38.202 (services provided by the physical layer of NR). These specifications also detail 5G protocol architecture and functional split of it, and the overall and radio protocol architectures which include Medium Access Control (MAC), Radio Link Control (RLC), Packet Data Convergence Protocol (PDCP) and Radio Resource Control (RRC).

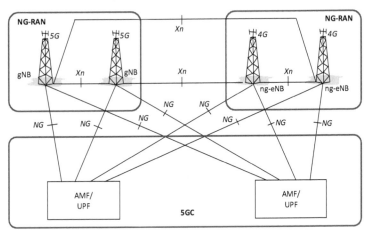

Figure 25 The 5G radio network architecture

The new radio interface covers the interface between the UE and the network on Layer 1, 2 and 3. The TS 38.200 series describe the Layer 1 (Physical Layer) specifications. Layers 2 and 3 are described in the TS 38.300 series.

Further reading	
▶3GPP	*TS 38.300* (New Radio)
▶5G Explained	Chapter 5 (radio network)

Requirements

While 3GPP makes the concrete technical 5G specifications, the International Telecommunications Union (ITU) Recommendation ITU-R M.2083 sets the reference presenting requirements for 5G, defining the IMT-2020 overall aspects and enhanced capabilities.

The IMT-2020 (International Mobile Telecommunications 2020) supports many deployment scenarios for different environments and service capabilities. ITU foresees 5G as an enabler for a seamlessly connected society as of the year 2020. The idea of 5G is to connect people via a set of "things", data, applications, transport systems and cities in a smart networked communications environment.

The idea of 5G was presented as early as in 2012 when ITU-R initiated a programme to develop International Mobile Telecommunications for the year of 2020 and beyond. The ITU-R working party 5D has been paving the way for the IMT-2020, including the investigation of the key elements of 5G in cooperation with the mobile broadband industry and other stakeholders interested in the 5G.

ITU-R's 5G vision for the mobile broadband connected society was agreed in 2015. This vision is considered as instrumental and it serves as a solid foundation for the World Radiocommunication Conference 2019 (WRC-19) which decides the international allocation of additional 5G frequency spectrum.

The ITU-R Rec. M.2083 binds multiple documents which forms the 5G. This publication includes IMT Vision (framework and overall objectives of the future development of IMT for 2020 and beyond), future technology trends of terrestrial IMT systems during 2015-2020, technical feasibility of IMT in bands above 6 GHz, framework and overall objectives of the future development of IMT-2000 and systems beyond IMT-2000, among others.

Table 11 Key requirements of the ITU IMT-2020, based on the M.2083.

Attribute	Value	Examples (scenario-dependent)
Dense areas	1 million devices / km^2	eMBB, sensor networks, ad-hoc broadband, massive IoT
Real time latency	Down to 1 ms	Tactile Internet, industry use cases, critical communications; According to TR 38.913, the NR should support latencies down to 0.5 ms UL/DL for URLLC
User mobility	Up to 500 km/h	Very fast-moving vehicles such as bullet trains
Throughput	10-20 Gb/s/user	Mobile broadband; in eMBB, target max for peak data rate is 20 Gb/s in DL and 10 Gb/s in UL
Ultra-high reliability	Up to 99.999 %	Remote surgery, remote control of objects such as drones, lifeline communications
Ultra-low cost	Sub-10 usd devices	Low-cost IoT, data offloading, network function virtualization, network slicing
Energy efficiency	Battery life up to 10 years	Considerably longer battery life than in previous generations

Further reading	
▶ITU	*ITU-R IMT-2020; ITU-R M.2083* (IMT Vision of 5G)
▶5G Explained	Section 2.3 (5G requirements based on ITU)

S

Security

5G has been standardized with a special emphasis on the integrated and enhanced security from day one.

While several security requirements such as the need for identification, mutual authentication, confidentiality, integrity and privacy protection remain unchanged compared to existing technologies, the expected new 5G use cases have had important impact on its renewed security architecture.

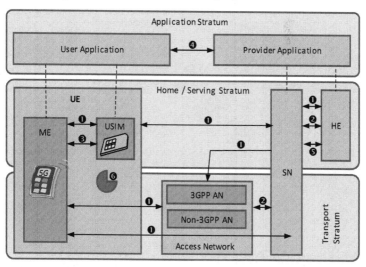

Figure 26 3GPP system security architecture for 5G.

The 3GPP 5G system consists of evolved physical radio and core networks and their logical functions. The security architecture can be presented as security domains that consist of Application Stratum (AS), Home Stratum (HS), Serving Stratum (SS) and Transport Stratum (TS). Figure 26 outlines the functional elements within these Stratums, and their security relations. Table 12 summarizes the interfaces of Figure 26.

Table 12 The interfaces of 3GPP security architecture.

Interface	Function
1	Network access security features for UE to authenticate and access services securely via network. These features protect the radio interface, whether it is 3GPP or non-3GPP radio access, and deliver security context of Serving Network (SN) to UE.
2	Network domain security features provide means for network nodes to securely exchange user data and signaling.
3	User domain security features secure the user access to ME.
4	Application domain security features provide means for applications of user and provider to exchange messages securely.
5	Service-Based Architecture (SBA) domain security features include network element registration, discovery, authorization security, and protection for the service-based interfaces.
6	Visibility and configurability of security features provide means to inform user if a security feature is in operation.

Figure 27 depicts the 5G network security functions in non-roaming scenario. In addition to these security interfaces, 3GPP also defines a Security Edge Protection Proxy (SEPP) which is designed to protect the messages of the *N32* interface.

AUSF: The *Authentication Server Function* replaces the MME / AAA of the 4G system. It terminates requests from the SEAF and interacts with the ARPF. The AUSF and the ARPF could be collocated and form a general EAP server for EAP-AKA and EAP-AKA'.

ARPF: The *Authentication Credential Repository and Processing Function* is collocated with the UDM (Unified Data Management). It stores the long-term security credentials such as user's key *K*. Based on those, it executes cryptographic algorithms, and it creates authentication vectors.

SCMF: The *Security Context Management Function* retrieves the key from the SEAF, which is used to derive further keys. The SCMF may be collocated with the SEAF in the same AMF (Access and Mobility Management Function).

SIDF: The *Subscription Identifier De-Concealing Function* is a service offered by the UDM network function of the home network of the subscriber. It de-conceals the SUPI (Subscription Permanent Identifier) from the SUCI (Subscriber Concealed Identifier).

SEAF: The *Security Anchor Function* forms, as a result of the primary authentication, the unified, common anchor key K_{SEAF} for all the access scenarios. K_{SEAF} protects the communications of the UE and the serving network, and it resides in the visited network in roaming scenario. There may be separate K_{SEAF} keys for the same UE connected to a 3GPP and a non-3GPP (such as Wi-Fi) access networks. SEAF is collocated with the AMF in 3GPP Release 15.

SPCF: The *Security Policy Control Function* provides policies related to the security of network functions such as AMF, SMF and UE. The elements involved for each policy scenario are dictated by the AF. The SPCF may be collocated with the PCF, or it can be a stand-alone element. SPCF contributes to the confidentiality and integrity protection algorithms, key lifetime and length, and the selection of the AUSF.

Thanks to the mobile edge computing and virtualization, 5G network functions and contents are moving closer to the consumer. The functions and contents are often replicated and exposed in potentially less protected environments. To make contents available to the user with reduced latency, the edges need to cache it via 3rd party content provider, so there needs to be adequate security measures in place, respectively.

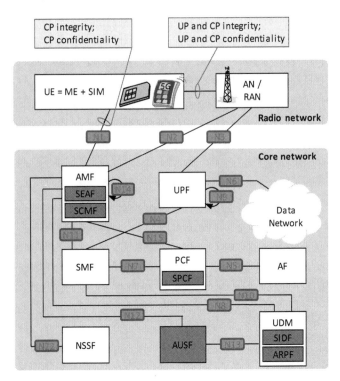

Figure 27 5G network functions related to security (highlighted).

Authentication between network entities ensures that the edge is authorized to receive the NF instance or the content. Also, the data being exchanged between those entities need to be protected at rest and in transit. The security mechanisms in the edges will be based on virtualization technologies involving hypervisors and isolation.

Further reading	
▶3GPP	*TS 33.401* (LTE security); *TS 33.501* (5G security)
▶5G Explained	Chapter 8 (security)

SIM in 5G Era

5G will still rely on the "traditional" SIM cards. Nevertheless, there will be many other forms such as embedded and integrated variants in the consumer and IoT markets of 5G.

The removable UICC (Universal Integrated Circuit Card), commonly known as SIM card (Subscriber Identity Module), is based on ISO/IEC 7816 smart card standard. It was adopted to GSM and other 3GPP systems, and is used nowadays in some non-3GPP systems, too.

It will maintain the relevance as a tamper-resistant, hardware-based secure element to store keys and other confidential information also in 5G. This is logical as there is an important base of legacy devices and established UICC ecosystem for the manufacturing, logistics and personalization of the operators' UICC profiles. The UICC provides adequate security level compliant with the strict international requirements of GSMA Security Accreditation Scheme (SAS) and payment institutions.

All the currently utilized ETSI-specified form factors (FF) as depicted in Figure 28 (size in mm) are physically compatible with 5G although some of the previous UICC hardware variants cannot necessarily handle the increased processing of 5G. 5G needs merely some changes to the UICC's contents (file structure) as there are new files defined for 5G functionalities in the Standalone mode.

With fast development of IoT and the need for respective built-in security, the market will benefit from a deeper integration of the USIM (Universal Subscriber Identity Module forming file structure within UICC) functionality into these devices. An ultra-small device would not be able to house a traditional UICC form factors. Thus, 5G markets are expected to use smaller UICC form factors of embedded and integrated UICCs. New over-the-air provisioning technologies are standardized, too, to support the embedded and integrated variants as they are soldered permanently into the device.

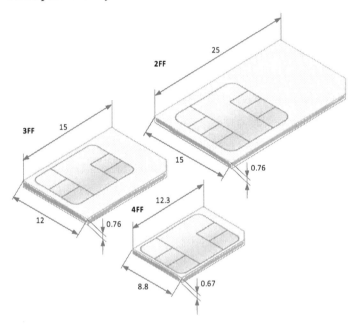

Figure 28 The "traditional" UICC (SIM card) form factors are also valid in 5G era.

The new interoperable variants of the UICC are facilitated by ETSI (European Telecommunications Standardisation Institute), which has widened the traditional definition of a UICC. ETSI refers the evolved UICC to as Smart Secure Platform

(SSP). Apart from this new model, the old UICC continues to exist while SSP is expected to be gradually more popular.

Meanwhile, also the Machine-to Machine Form Factor (MFF2) of ETSI will be still valid in 5G together with any proprietary variants of embedded UICC (eUICC), that is, permanently soldered UICCs into device. Figure 29 summarizes some examples of these options.

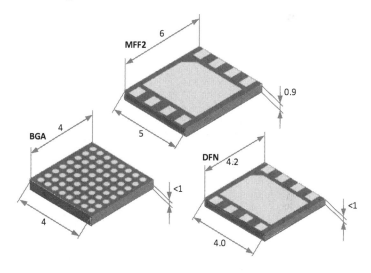

Figure 29 Some examples of embedded UICC elements.

For both embedded and integrated UICC, referred to as eSSP and iSSP, respectively, according to the new ETSI terminology, GSMA has developed eSIM specifications for the new way to manage the MNO profiles. Commonly known as Remote SIM Provisioning (RSP), the new eSIM ecosystem is being developed jointly with GlobalPlatform and SIMalliance.

3GPP defines the new 5G subscription credentials as a set of values in the USIM and the Authentication Credential Repository and Processing Function (ARPF). These credentials refer to a long-term key or set of keys K, unique for each user, and the Subscription Permanent Identifier (SUPI) which uniquely identifies a subscription. The *K* and SUPI are designed to mutually authenticate the User Equipment (UE) with the 5G core network.

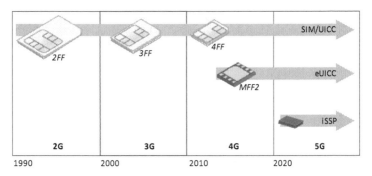

Figure 30 UICC variants and respective commercial timelines.

The subscription credentials are processed and stored in the 3GPP-defined USIM. In addition, 3GPP has confirmed the USIM still resides on UICC in the 5G era. In general, the same principles apply as previously defined for both the file structure and actual files throughout prior 3GPP generations.

The eSIM principles designed for the UICC profile management apply to 5G, too, and the evolved SIM can serve as a security anchor also to other stakeholders such as device manufacturers and application providers requiring security in the end-points. Moreover, the evolution of 5G subscription management is planned to have a convergence path to support both consumer and M2M devices by a common platform.

Further reading	
▶3GPP	*TS 33.501* (5G security architecture and procedures)
▶GSMA	*GSMA eSIM*. (GSMA, 2019)
▶5G Explained	Section 8.10 (UICC evolution)

Specifications

The initial 5G is defined as of the 3GPP Release 15 specifications. After the first phase, 5G, as described in Release 16, will comply with the demanding ITU IMT-2020 requirements.

The original GSM (Global System for Mobile Communications) was deployed commercially 1991 based on the ETSI (European Telecommunications Standardisation Institute) specifications. ETSI continued producing GSM releases until the 3GPP (3rd Generation Partnership Project) was formed as a successor for carrying out the work of ETSI in 1999.

Nowadays, 3GPP maintains and enhances the specifications for GSM, UMTS/HSPA, LTE/LTE-A and 5G. 3GPP allows free access to the respective specification at www.3gpp.org. The status of each specification can be found at the 3GPP web page:

https://www.3gpp.org/DynaReport/status-report.htm

This link is one of the ways to find the Technical Specifications (TS) and Technical Reports (TR) per each 3GPP Release. As an example, the first phase 5G New Radio (NR) user equipment description is found in the Release 15 section of the above-mentioned page by downloading the TS 38.306 (NR; User Equipment radio access capabilities).

While 3GPP produces the concrete specifications for 5G, there are other standard development organizations (SDOs) and industry forums contributing to the ecosystem of the 5G. One of such entities is the GSMA which, among other tasks, works with the member organizations on the interoperability of the networks and develops services on top of the 3GPP definitions. Some examples are VoLTE (Voice over LTE) and RCS (Rich Communications Services) which will be valid also in 5G era in Non-Standalone deployments, and in a renewed form of VoNR (Voice over New Radio) in the Standalone networks.

Table 13 and Table 14 summarize some 5G documents of the 3GPP and GSMA. These documents include further references for more detailed exploration of 5G.

Table 13 Some key 5G specifications of the 3GPP.

TS/TR	Title	Area
TR 22.891	Feasibility study on new services and market technology enablers	Services
TS 23.501	System architecture of the 5G system	Core
TS 23.502	Procedures for the 5G system	Functionality
TR 25.903	Deployment aspects	Network planning
TS 33.501	Security architecture	Security
TS 38.101	NR; User Equipment (UE) radio transmission and reception;	Radio
TS 38.104	NR; Base Station (BS) radio transmission and reception	Radio
TS 38.201	NR, general description	Radio

Table 14 5G-related Permanent Reference Documents of the GSMA

PRD	Title	Area
NG.116	Generic 5G Network Slice Template	Network slicing

Further reading	
▶3GPP	*5G specifications of Release 15*
▶5G Explained	Section 2.4 (the technical specifications of 3GPP)

Standardization

While ITU sets the overall requirements for 5G, 3GPP is the "work horse" which makes the concrete standards.

ITU-R (radio section of the ITU) has defined universal requirements and principles for global 5G under the term IMT-2020 (International Mobile Telecommunications 2020). Figure 31 depicts the principle of the liaison between ITU, 3GPP, and other forums.

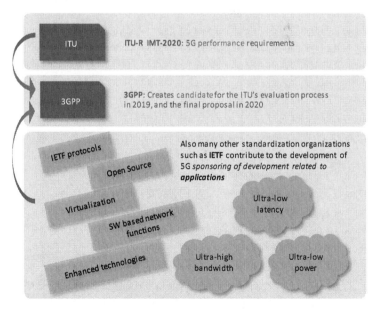

Figure 31 The main 5G standardization body is the 3GPP.

3GPP 5G technical specifications are divided into first phase as per Release 15, and second phase as per Release 16. The Release 17 is already under planning, representing evolved second phase together with the forthcoming Releases after that.

The Release 15 does not quite yet provide with sufficient performance for complying with the IMT-2020 requirements. Release 16 is planned to do so as defines the full version of 5G.

There are also other Standard Development Organizations (SDOs) and industry forums contributing to the development of the 5G ecosystem. Table 15 lists some of the relevant ones.

Table 15 Some key entities standardizing and evolving 5G.

Entity	Activities in 5G
3GPP / 3GPP2	Produces 5G technical specifications as of Release 15
5GAA	Promotes 5G in automotive ecosystem
C2C CC	Car-to-Car Communication Consortium, industry forum for the V2V technology development
CSA	Cloud Security Alliance promotes security assurance
EITF	Internet architecture; protocols adapted in 5G
ETSI	SIM evolution (SSP); integrated SSP for 5G
Global Standards Collaboration	Enhances global standards. Members: ARIB, ATIS, CCSA, ETSI, IEX, IEEE-SA, ISO, ITU, TIA, TDSI, TTA, TTC
GlobalPlatform	UICC development
GSMA	Remote SIM provisioning; Voice and messaging services over 5G; Network slicing interoperability
IEEE	IEEE 802 series; IoT standards
ISO	IT security; Smart Card standardization; Common Criteria
ITU	Requirements of 5G in form of IMT-2020
NGMN	Next Generation Mobile Networks, optimizes and guides on advanced network technologies and IoT
NIST	Cybersecurity framework
OMA	Device management; LightweightM2M
SIMalliance	Secure Element implementation; eUICC, iUICC
STA	Secure Technology Alliance Facilitates the adoption of secure solutions in the U.S.

Further reading	
▶3GPP	*3GPP web page* (http://www.3gpp.org/)
▶5G Explained	Section 1.9 (5G standardization and regulation)

Terminal States

The 5G User Equipment must perform a network registration and authorization prior to using services.

When 5G device is switched on, it performs initial network attach procedure (registration). Other signaling flows include call establishment and mobility management procedures in the idle and connected modes, PDU (Packet Data Unit) session establishment (5G data connection), session termination, and network deregistration. 5G performs detach procedure when the user switches off the equipment, or it runs out of the battery.

Figure 32 and Figure 33 depict the state diagram of 4G and 5G, and the 3GPP TS 23.502 details both of them.

The registration procedure happens when the UE performs the initial registration to the 5G system, or when it executes Mobility Registration Update procedure whenever the Tracking Area (TA) changes. Tracking area is a set of gNBs. When 5G customer is about to receive a connection request, the network sends paging signaling to the UE within the TA where it was last registered via all the gNB elements.

If the TA is large, it triggers big amount of radio signaling. The UE needs to send registration signaling to network each time it moves between the TA areas. If the TA areas are dimensioned very small, it increases signaling and consumes the UE's battery, so it is important to optimize the TA size.

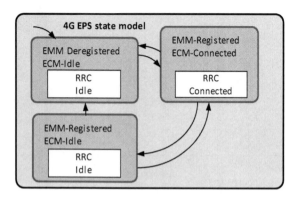

Figure 32 The state model for 4G system.

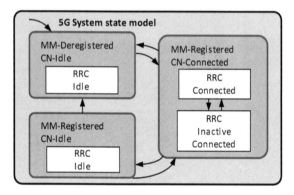

Figure 33 The state model for 5G system.

The registration is also done to update the UE capabilities, and in a Periodic Registration Update. In 5G deregistration procedure, the UE informs the network to stop accessing the 5G core, and the 5G core informs the UE about the released access.

Further reading	
▶3GPP	*TS 23.502* (procedures for the 5G system)
▶5G Explained	Section 7.2.2 (5G states)

User Equipment

There will be many familiar looking as well as totally new type of mobile devices in the 5G era. These devices will support a variety of usage types including smart devices and fixed wireless access via CPE (customer premise equipment).

There will also be IoT devices such as Machine-to-Machine equipment, intelligent sensors, and modules for automotive and industry, as well as new, advanced devices for Virtual Reality.

The Technical Specification 3GPP TS 38.306 defines the 5G User Equipment (UE) radio access capabilities as of Release 15, including new terminal aspects, local connectivity, evolved battery, display, sensor, memory and processor technologies. Table 16 summarizes other relevant specifications for the UE.

The 3GPP TS 38.300 defines the UE capabilities for 5G New Radio (NR). In practice, the network determines the uplink (UL) and downlink (DL) data speed of UE based on the supported band combinations and baseband capabilities, i.e., modulation scheme, multiple-in, multiple-out (MIMO) antenna layers, and other characteristics on the radio path.

According to GSA (Global Mobile Suppliers Association), there were early-stage device types in the consumer markets by 19 vendors in July 2019, including phones, hotspots, customer premise equipment, modules, "Snap-On" dongles and adapters, and USB terminals, while the 5G chipset modules are becoming to be available. (GSA, n.d.)

In critical communications, 5G devices will be able to communicate directly with other devices – similarly as is the case with the push-to-talk service already today. This is beneficial for special circumstances as, e.g., police and fire brigades require reliable radio communications in emergency areas even when there is no network coverage available. The respective security requirements include mutual authentication of a single or a group of devices without need to negotiate and agree keys, as well as data encryption and integrity protection.

Table 16 Some of the key 5G specifications describing the UE.

3GPP TS	Title
38.300	UE key functionalities and capabilities
38.101-1	NR User Equipment (UE) radio transmission and reception, Part 1, Range 1 Standalone
38.101-2	NR User Equipment (UE) radio transmission and reception, Part 2, Range 2 Standalone
38.101-3	NR User Equipment (UE) radio transmission and reception, Part 3, Range 1 and range 2 interworking operation with other radios
38.101-4	NR User Equipment (UE) radio transmission and reception, Part 4, Performance requirements

Security-wise, only the authorized users may access the 5G device. This can be achieved by verifying a PIN or biometric characteristic of the user. In addition, theft prevention should be considered so that only authorized entities may be able to disable or re-enable stolen devices.

Also, users may want to store and process private and confidential data on the device, leveraging the secure entity within the device for this purpose.

Further reading	
▶3GPP	*TS 38.300* (New Radio)
▶5G Explained	Section 5.4 (5G radio access technologies)

V2V

One of the important use cases of 5G is related to Vehicle-to-Vehicle (V2V), or more generally, Cellular-based Vehicle-to-Everything (C-V2X).

V2V refers to the direct links between vehicles. Vehicle to Infrastructure (V2I) involves other components such as roadside elements. Vehicle to Pedestrian (V2P) means communication between vehicles and nearby persons, while V2N refers to Vehicle-to-Network communications. V2X consists a combination of V2N, V2V, V2I and V2P.

The strict telecommunication requirements needed for automotive environment, including the self-driving vehicles, map with the full version of 5G performance and capabilities.

The previous V2V system is a Wi-Fi -based IEEE 802.11p. Figure 34 depicts its architecture. It is designed to form fast, adaptive ad-hoc connectivity between vehicles. The vehicles also communicate with the Road Side Units (RSU) which connect external networks. The roadside units act as communicating nodes relaying information such as safety warnings and traffic information. The positive aspects of this technology include established technology while the sufficiently large coverage of the RSUs may be expensive.

On July 2019, European Union is forming guidelines for the application of IEEE 802.11p or 5G technology for European automotive markets. Even 5G is still less mature than the Wi-Fi -based system, 5GAA (5G Automotive Association) has identified 5G technology as an important alternative. The benefits of 5G include integrated security and the reuse of the radio and core infrastructure which optimizes the access to external networks. 5G. (Chee, 2019)

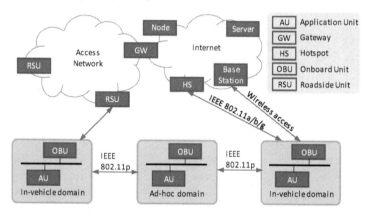

Figure 34 The principle of the IEEE 802.11p architecture.

According to 5GAA, C-V2X does not necessarily require network infrastructure. It can also operate without SIM or network assistance and use satellite positioning (GNSS) as its primary source of time synchronization.

C-V2X is defined under the term LTE V2X in the 3GPP Release 14, and is designed to operate in several modes, including device-to-device operation. (5GAA, n.d.)

Although the current downside of 5G is the lack of coverage, the C-V2X can operate already via LTE. As the 5G matures, it can assume more important role for the V2X communications.

Further reading	
▶3GPP	*TR 22.885* (LTE support for Vehicle-to-Everything)
▶5G Explained	Section 3.4 (The role of 5G in automotive)

Virtual Reality (AR/VR/XR)

The enhanced performance of 5G will benefit many new applications that were impossible to use with the legacy mobile communication systems. Some of the new applications benefiting from 5G is related to Augmented and Virtual Reality.

The increasing number of modern services and applications benefit from enhanced network performance of 5G to keep providing the users with the most fluent experiences. Only time will show how fast 2G, 3G, and 4G will lose consumers' interest as the more capable and spectral efficient 5G will take over. One of the indications of the forthcoming popularity of 5G can be seen from the recent GSMA study which forecasts that 5G is to account for 15 percent of global industry by 2025.

The 3GPP Release 16 technical specification set represents an important step in this evolution. It provides the mobile network operators with the means to start deploying a further evolved, fully IMT-2020 -compatible 5G as of mid-2020. Those networks will be capable of providing the lowest latencies and highest data speeds, among other enhancements.

Along with the commercial Release 16 networks, 5G will open many new business opportunities for the established and completely new stakeholders over the whole ecosystem, including application developer community representing highly advanced technologies. One example of the expected new applications benefiting from the 5G performance and capacity is related to the augmented reality (AR) and virtual reality (VR). In order for them to work fluently, they require very low latency for communications and data processing.

Typically, prior to 5G, such applications have been relying on a high-performance processing and expensive stand-alone hardware located at site. As the mobile 5G is based on network functions virtualization, high speed data, and edge computing, the applications could offload the device's data processing onto the cloud. The AR/VR applications, such as multi-player video games with a complex modelling of avatars, are especially interesting area as the user actions require nearly real-time responsiveness from the devices and infrastructure. According to (GSMA, 2018), the AR/VR is expected to be a disruptive form of immersive multimedia. It will also be applied to a diverse range of services and use cases and change the way new content is delivered.

By applying AR/VR data rendering to offload data processing from the user device onto the edge cloud, it would be also possible to reduce the handset processing requirements and cost. This opportunity opens highly relevant and new co-operative business models for mobile network operators designing advanced added-value services, which were not feasible in the older mobile communication infrastructure.

GSMA is investigating and promoting the 5G cloud technology for the VR/AR applications via the recently established Cloud XR Forum. (GSMA, 2018) The aim of the forum is to accelerate the delivery and adoption of 5G cloud-based AR and VR technologies.

Further reading

▶**GSMA** *Future Networks Program, Cloud XR Forum*

Voice Calls

5G is a packet switched system and does not thus include circuit switched connectivity. The native IP voice call service of 5G is referred to as VoNR (Voice over New Radio).

The initial 5G deployment is typically based on the Option 3 (see Deployment Chapter). It gives mobile network operators a possibility to rely on their legacy 4G core networks by attaching gradually new radio elements of 5G.

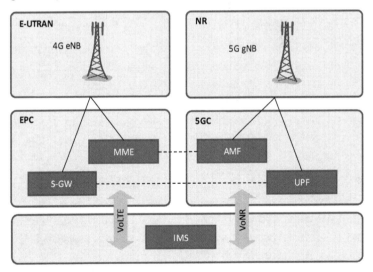

Figure 35 The interconnection of 4G and 5G for the VoLTE and VoNR.

Meanwhile, the respective 4G platforms and services like IMS (IP Multimedia Sub-system) and VoLTE (Voice over LTE) remain largely valid for 5G users, either as such or with only minor modifications.

In the initial phase of 5G, IMS is not able to distinguish between the 4G eNB and 5G gNB. In the initial deployment, the 5G voice service relies still on the 4G EPC and VoLTE to facilitate a seamless voice call experience thanks to the fallback mechanisms. Both 4G and 5G radio, core, and IMS networks can support the voice service and the respective capabilities.

In the final SA deployment phase, based on the Option 2, the native 5G VoNR serves the New Radio voice call users with call continuity from the VoNR back to previous systems.

The VoLTE and VoNR deployment can be done in phases:

1) **5G NSA**: 5G may rely on the VoLTE service;

2) **5G SA,** *first phase*: For the voice calls through VoLTE, Evolved Packet System Fallback (EPS FB) can be used;

3) **5G SA,** *second phase*: As the number of devises supporting the SA mode increases, VoNR can be selected as a default voice service. If the user moves from the 5G to 4G radio coverage area during the call, a packet switched handover takes care of the seamless voice service continuity from the 5G NR to 4G VoLTE. This procedure is comparable with the SRVCC (Single Radio Voice Call Continuity) of 4G.

According to GSMA, as soon as the 5G radio and core networks are deployed widely enough facilitating a continuous service, the IMS voice and video calls via the 5G infrastructure are superior as for capacity compared to the same services on LTE. (GSMA, 2018)

Further reading	
▶3GPP	*TS 23.501* (Section 4.4.3); *TS 23.502* (Section 4.13.6)
▶GSMA	*Road to 5G: Introduction and Migration* (White Paper)

World Radiocommunication Conference

International Telecommunications Union (ITU) organizes World Radiocommunication Conferences (WRC) to discuss and decide global allocations for the radio frequency bands. The work aligns the efforts of international frequency administrations. The most relevant topic for the forthcoming WRC-19 is related to 5G bands.

The WRC events are organized every three to four years. The latest WRC was organized in 2015, and the 2019 event is important for the further 5G frequency allocations. WRC-19 will take place 28 October to 22 November 2019. (ITU, 2019)

WRC reviews and revises global radio regulations. The international treaty governs the use of the radio frequency spectrum as well as the geostationary satellite and non-geostationary satellite orbits.

During the preparation of each WRC, the ITU Council takes into account recommendations of the previous world radiocommunication conferences, and coordinates the meeting for the member states. WRC then revises the radio regulations and associated frequency assignment and allotment plans. The Conference Preparatory Meeting (CPM) will prepare a consolidated report to be used in support of the work of the conferences. (ITU, n.d.) The world and regional Radiocommunication Conferences consider contributions from administrations, the Radiocommunication Study Groups, and other sources related to the regulatory, technical, operational and procedural matters.

WRC is a formal forum for setting general rules for 5G and other radiocommunication frequency allocations, and is highly relevant for supporting 5G deployment strategies. The allocation plans are also the most important base for forthcoming 5G frequency auctions.

The ITU regions are divided into 3 main zones: Asia, Europe and Americas regions. The overall allocation plan differs from region to region. For 5G and previous generations, there are both global and regional bands to be considered by regional authorities. As a consequence, in order to be a global variant, 5G User Equipment needs to support a common set of internationally compatible frequencies from all the three regions.

In 5G, the task is not straightforward as the support of multiple bands increases the complexity, cost, and physical dimensions of the device. The optimization of an adequate set of terminal's frequencies is thus one of the important optimization tasks of the device manufacturers, and the outcome of the WRC has its impact on the future planning.

The next frequency allocation plan is one of the essential steps to understand the opportunities for future 5G bands. Nevertheless, along with the new allocation plans for the primary and secondary users, it is not always straightforward in practice to liberate old bands. They might have been serving still important applications, and their removal might not be always possible in the near future. The national regulators are in key position to balance the needs of old and new users.

Further reading	
▶ITU	*WRC web page* (https://www.itu.int/en/ITU-R/conferences/wrc/Pages/default.aspx)
▶5G Explained	5.3.2 (ITU-R WRC-19 expectations)

XHaul

The dimensioning of the transport network is one of the essential tasks of 5G operators. The fast development of the mobile communication systems has resulted in largely virtualized network architecture models and respective cloud infrastructure which can be used in core and radio systems. The transport network infrastructure develops accordingly to ensure adequate end-to-end performance.

5G XHaul is originally a European-wide project for converged Fronthaul (FH) and Backhaul (BH) networks of 5G. It describes a logical transport architecture integrating multiple wireless and optical technologies under a common Software Defined Networking (SDN) control plane. (5G PPP, 2016) 5G XHaul refers to a unified transport supporting split concept of flexible RAN (Radio Access Network). EU is planning to use XHaul as a base for the 5G transport infrastructure.

There is a unified SDN XHaul **control plane** designed for wireless and optical networks. As explained in Chapter Cloud RAN and Core on the respective transport control, also XHaul is aware of the variations in the RAN. Figure 36 depicts the principle of the XHaul control.

The **data plane** of XHaul unifies the transport of wireless networks such as P2MP (Point-to-Multipoint) mm-Wave radio, sub-6 GHz mobile networks, optical networks, Time Shared Optical Network (TSON), and Wavelength Division Multiplexing (WDM) -based Passive Optical Network (PON) referring to optical fiber network architecture. Figure 37 depicts the user plane of XHaul.

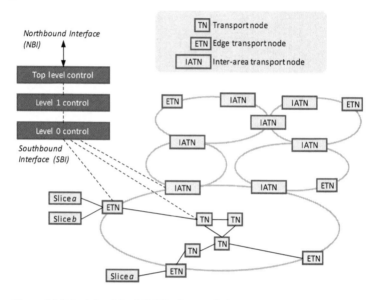

Figure 36 Principle of the 5G XHaul control plane.

These networks tolerate much greater data loads than the older systems were able to, and can provide wider bandwidth compared to the copper-based legacy networks.

XHaul serves as an example of some of the practical options and respective performance and functional requirements for 5G transport network data transfer. More up-to-date information on the architecture of the XHaul is available at the web page of 5G PPP. (5G-XHaul, n.d.)

The XHaul thus refers to a new RAN transport infrastructure which supports fronthaul, mid-haul and backhaul. Its physical layer is based on fiber optics, metro SDN and edge cloud solutions for serving 5G signaling and data traffic.

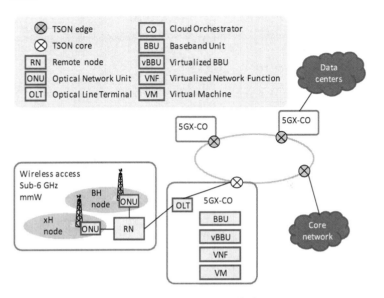

Figure 37 Example of the 5G-XHaul user plane deployment.

As stated in (Brown, 2019), there are various requirements for the Open XHaul in 5G environment, including the following:

- **Bandwidth**: 5G requires interfaces which are capable of supporting 100 Gb/s data speeds for radio access networks based on cost-effective fiber optics.

- **Interoperability**: The 5G fronthaul needs to support a combination of legacy protocols such as CPRI, and new ones such as evolved CPRI (eCPRI) and Time Sensitive Networking (TSN).

- **Latency**: The round-trip time for many 5G applications need to be in range of sub-10 ms whereas the most demanding applications require latency values of down to 1 ms. The fronthaul schemes may require even faster responses as stated in IEEE 1914.3.

- **Synchronization**: 5G brings along with strengthened requirements for Time Division Multiplex (TDD)-based synchronization.

Further reading	
▶5G PPP	*View on 5G Architecture*, 5G PPP Architecture Working Group (5G PPP, 2016)
▶5G Explained	Section 9.2.2 (transport)

5G TERMINOLOGY

This section presents some of the most common terms related to 5G.

3GPP (3rd Generation Partnership Project) is a standards development organization (SDO) defining the technical specifications of 5G.

5G refers to the fifth generation of mobile communication systems. The formal criteria for mobile communication systems, in order to be called "5G", are set by ITU (International Telecommunications Union) and documented in IMT-2020 (International Mobile Telecommunications) requirement set.

Band is a generic term for the proportion of radio frequency. Each 5G operator has one or more frequency bands which are used for radio transmission and reception. Operators typically purchase the right to use frequency bands for their operations from auctions organized by national regulators. In 5G, the low-band refers to spectrum below 1 GHz, mid-band to 1-6 GHz, and high-band to spectrum range above 6 GHz.

Base Station refers to a physical site where radio and transmission equipment are located for interconnecting user equipment (UE) and cellular network. In 5G, the radio component is called gNB (next generation Node B). In LTE, it is eNB (evolved Node B), and in 3G it is NB (Node B).

CA (Carrier Aggregation) refers to the combining of radio capacity of two or more radio frequency bands for a single user.

CA has been applied since 2G, but 5G is able to benefit even more from it, thanks to the wider defined bandwidths.

Core Network connects the 5G base stations and delivers the data traffic and signaling between the user equipment and destination such as other mobile user equipment or server. In 5G, the core network is based on virtualized architecture, clouds, and data centers.

DSS, Dynamic Spectrum Switching, automatically allocates capacity for the 5G radio network by sharing the operator's same band with 4G frequencies as per need.

EMF (Electro Magnetic Field) is a common term referring to the radiation properties of radio transmitters, and it is related to safety regulation and health aspects. While EMF refers to radio waves influencing living tissues, the **EMC** (Electromagnetic Compatibility) is applicable term for indicating the influence of radio waves to equipment and electronics.

eSIM is electronic version of the traditional SIM card. It means that the functionality of the SIM card is embedded into the user equipment such as smart device. As some of the special devices such as sports trackers are small, the removable SIM might not fit into it. Thus, there are new SIM variants which can be soldered and integrated into the hardware of the device. Furthermore, there are new solutions to manage the subscription over the air instead of switching the SIM between devices. RSP (Remote SIM Provisioning) of GSMA is one of these methods.

Licensed band refers to the radio frequencies operators pay for to get right to use them for their customers. National regulators typically auction licenses to be used for limited time period.

MIMO refers to Multiple In, Multiple Out antenna type which makes it possible to transmit and receive via multiple radio paths enhancing the data speed.

Millimeter Wave, or mm-Wave, is the term for the new frequency bands on much higher frequencies than were deployed in previous systems. 3GPP defines Frequency Range 2 for mm-Wave bands whereas Frequency Range 1 is for sub-6 GHz.

NR refers to New Radio which is the new 3GPP standard for 5G radio system. It enhances the radio performance providing faster data speeds, more capacity and lower delay (latency).

NSA, Non-Standalone 5G network, is based on intermediate 5G architecture model which reuses part of the legacy 4G infrastructure. By using NSA in the initial phase, operator can expedite the offering of 5G services while constructing 5G core.

OFDM, Orthogonal Frequency Division Multiplexing, serves as 5G access technology. It is familiar from the radio interface of LTE and Wi-Fi, and is adjusted to work optimally in 5G radio with varying conditions. It has multiple sub-carriers so in case part of the transmission is lost, e.g., due to fading on certain frequency, the remaining sub-carriers can deliver the contents.

Re-farming means reallocation of a part or complete set of frequency bands from one to another mobile communication system to balance the offered capacity in an optimal way.

SA, Standalone 5G network, is the full architectural version of 5G, including its own new radio and core networks. SA is the ultimate goal of operators as it is able to offer the maximum performance without the limitations of legacy 4G.

Spectrum Sharing refers to techniques to optimize the radio network capacity by letting more than one provider to offer the same band for their customers. 4G has already solutions such as SAS (Spectrum Access System) and US-based CBRS (Citizens Broadband Radio Service). There is also further optimized method introduced in 5G called Dynamic Spectrum Sharing (DSS), which changes automatically the capacity from the same band among 4G and 5G users, and LSA (Licensed Shared Access).

Terminal Equipment (TE) refers to the sole hardware and software of 5G device without SIM (UICC+USIM). The user can establish only emergency calls via 5G with pure TE. The SIM is needed for voice and data connectivity. The combination of TE and SIM is referred to as User Equipment (UE).

Unlicensed band does not require licensing fees, and they can be shared among many providers. The downside is that the provider of such bands cannot guarantee the service level because of varying capacity demands of users. Examples of unlicensed band solutions include CBRS (Citizens Broadband Radio Service), LAA (Licensed Assisted Access), and NR-U (New Radio of 5G in Unlicensed mode as of Release 16).

UICC refers to a Universal Integrated Circuit Card that is a secure element (SE). In practice, it is called SIM in general. USIM (Universal SIM) is an application residing in the physical UICC. The role of "traditional", removable SIM can be assumed to be less important in 5G era while the embedded and integrated variants will be more popular thanks to their smaller size, and the possibility to manage them over the air.

Use case refers to a specific situation in which a product or service could potentially be used. Moreover, a use case is a methodology in system analysis to identify, clarify, and organize system requirements.

User Equipment (UE) is a 5G device which consists of Mobile Equipment (ME), or Terminal Equipment (TE), and Universal Subscriber Identity Module (UICC). In practice, the latter is oftentimes referred to as SIM (Subscriber Identity Module). Without SIM, the sole 5G ME can only be used for emergency calls.

BIBLIOGRAPHY

3GPP. (2018, December 14). *RAN adjusts schedule for 2nd wave of 5G specifications*. Retrieved from 3GPP: https://www.3gpp.org/news-events/2005-ran_r16_schedule

3GPP. (2018, July 16). *Release 16*. Retrieved from https://www.3gpp.org/release-16

3GPP. (2018). *TS 33.501*. 3GPP.

3GPP. (2019, June 3). *Legal Matters*. Retrieved from Legal Status: https://www.3gpp.org/about-3gpp/legal-matters

5G Americas. (2019, February). *The status of Open Source for 5G*. (5G Americas) Retrieved July 7, 2019, from P4 is an open-source initiative designed primarily to provide a declarative language for interacting with networking forwarding planes

5G PPP. (2016, December 31). *5G-XHaul*. (5G PPP) Retrieved July 6, 2019, from D2.3 Architecture of Optical / Wireless Backhaul and Fronthaul and Evaluation: https://5g-ppp.eu/new-deliverables-available-for-5g-xhaul/

5GAA. (n.d.). *V2X*. (5GAA) Retrieved July 5, 2019, from Exploring the technology: C-V2X: http://5gaa.org/5g-technology/c-v2x/

5G-XHaul. (n.d.). (New deliverables available for 5G-XHaul) Retrieved July 7, 2019, from 5G PPP: https://5g-ppp.eu/new-deliverables-available-for-5g-xhaul/

Ann Armstrong, J. M. (2014, May 29). *The Smartphone Royalty Stack: Surveying Royalty Demands for the Components Within Modern Smartphones*. Retrieved from SSRN: https://papers.ssrn.com/sol3/papers.cfm?abstract_id=2443848

Bougioukos, M. (2017). *Preparing microwave transport network for the 5G world.* Nokia.

Brown, G. (2019). *New transport network architectures fro 5G RAN.* Retrieved from A Heavy Reading white paper produced for Fujitsu: https://www.fujitsu.com/us/Images/New-Transport-Network-Architectures-for-5G-RAN.pdf

Chee, F. Y. (2019, July 4). *EU opens road to 5G connected cars in boost to BMW, Qualcomm.* (Reuters) Retrieved July 5, 2019, from Reuters: https://www.reuters.com/article/us-eu-autos-tech/eu-opens-road-to-5g-connected-cars-in-boost-to-bmw-qualcomm-idUSKCN1TZ11F

Ericsson. (n.d.). *5G deployment considerations.* (Ericsson) Retrieved July 5, 2019, from Insights and reports: https://www.ericsson.com/en/networks/trending/insights-and-reports/5g-deployment-considerations

ETSI. (2018). *TS 133.501, V. 15.1.0, Chapter 5.2.5: Subscriber Privacy.* ETSI.

FCC. (2015, November 25). *RF Safety FAQ.* Retrieved from Radio Frequency Safety: https://www.fcc.gov/engineering-technology/electromagnetic-compatibility-division/radio-frequency-safety/faq/rf-safety

FCC. (2016, December 15). *Specific Absorption Rate (SAR) for Cellular Telephones.* (FCC) Retrieved July 4, 2019, from Federal Communications Commission: https://www.fcc.gov/general/specific-absorption-rate-sar-cellular-telephones

GSA. (n.d.). *5G Device Ecosystem Report March 2019.* (GSA) Retrieved July 4, 2019, from GSA: https://gsacom.com/paper/5g-device-ecosystem-report-march-2019/

GSMA. (2017). *iUICC Assurance report.* GSMA.

GSMA. (2018, November 29). *Future Networks.* (GSMA) Retrieved July 5, 2019, from GSMA Launches New Forum to Support 5G Cloud-Based AR/VR: https://www.gsma.com/futurenetworks/digest/gsma-launches-new-forum-to-support-5g-cloud-based-ar-vr/

GSMA. (2018, May 14). *Mobile IoT in the 5G Future – NB-IoT and LTE-M in the Context of 5G*. (GSMA) Retrieved July 5, 2019, from GSMA: https://www.gsma.com/iot/mobile-iot-5g-future/

GSMA. (2018, April). *Road to 5G: Introduction and Migration* . (GSMA) Retrieved July 5, 2019, from GSMA: https://www.gsma.com/futurenetworks/wp-content/uploads/2018/04/Road-to-5G-Introduction-and-Migration_FINAL.pdf

GSMA. (2018, March 26). *The 5G era in the US*. (GSMA) Retrieved July 3, 2019, from GSMA: https://www.gsma.com/publicpolicy/resources/the-5g-era-in-the-us

GSMA. (2019). *EMF and Health*. Retrieved from www.gsma.com/emf

GSMA. (2019, February 25). *New GSMA Study: 5G to Account for 15% of Global Mobile Industry by 2025 as 5G Network Launches Accelerate*. (GSMA) Retrieved 6 30, 2019, from https://www.gsma.com/newsroom/press-release/new-gsma-study-5g-to-account-for-15-of-global-mobile-industry-by-2025/

GSMA. (2019, July). *Safety of 5G Mobile Networks*. Retrieved from https://www.gsma.com/publicpolicy/wp-content/uploads/2019/06/GSMA_Safety-of-5G-Mobile-Networks_July-2019.pdf

GSMA. (2019, May 1). *South Korea already has over 260,000 5G subscribers*. (GSMA) Retrieved July 3, 2019, from GSMA: https://www.gsmarena.com/260000_subscribers_are_using_5g_in_south_korea_according_to_ministry_of_science_and_ict-news-36851.php

GSMA. (2019). *The SIM for the next Generation of Connected Consumer Devices*. Retrieved from eSIM: https://www.gsma.com/esim/

Health Canada. (2018, 18 April). *Limits of Human Exposure to Radiofrequency Electromagnetic Energy in the Frequency Range from 3 kHz to 300 GHz*. (Health Canada) Retrieved July 4, 2019, from Safety Code 6 (2015): https://www.canada.ca/en/health-canada/services/environmental-workplace-health/consultations/limits-human-exposure-radiofrequency-electromagnetic-energy-frequency-range-3-300.html

Horwitz, J. (2019, May 2). *China dusts the U.S., Finland, and South Korea with 34% of key 5G patents*. Retrieved from Venturebeat: https://venturebeat.com/2019/05/02/china-dusts-the-u-s-finland-and-south-korea-with-34-of-key-5g-patents/

IPlytics. (2019, July). *Who is leading the 5G patent race?* Retrieved from https://www.iplytics.com/wp-content/uploads/2019/01/Who-Leads-the-5G-Patent-Race_2019.pdf

ITU. (2018). *Focus Group on Technologies for Network 2030*. (ITU) Retrieved July 5, 2019, from FG NET-2030: https://www.itu.int/en/ITU-T/focusgroups/net2030/Pages/default.aspx

ITU. (2019). *Country ICT data (until 2017)*. (International Telecommunications Union) Retrieved July 5, 2019, from Statistics: https://www.itu.int/en/ITU-D/Statistics/Pages/stat/default.aspx

ITU. (2019). *World Radiocommunication Conference 2019 (WRC-19), Sharm el-Sheikh, Egypt, 28 October to 22 November 2019*. (ITU) Retrieved July 6, 2019, from WRC-19: https://www.itu.int/en/ITU-R/conferences/wrc/2019/Pages/default.aspx

ITU. (n.d.). *ITU*. (ITU) Retrieved July 6, 2019, from World Radiocommunication Conferences (WRC): https://www.itu.int/en/ITU-R/conferences/wrc/Pages/default.aspx

ITU-T. (2017). *ITU-T Focus Group IMT-2020 Deliverables*. Retrieved from https://www.itu.int/dms_pub/itu-t/opb/tut/T-TUT-IMT-2017-2020-PDF-E.pdf

Mobile World Live. (2019, June 13). *Operators scathing over German 5G auction outcome*. Retrieved from https://www.mobileworldlive.com/featured-content/home-banner/operators-scathing-over-german-5g-auction-outcome/

Murphy, K. (n.d.). *Centralized RAN and Fronthaul* . (Ericsson) Retrieved July 6, 2019, from https://www.isemag.com/wp-content/uploads/2016/01/C-RAN_and_Fronthaul_White_Paper.pdf

National Cancer Institute. (2019, January 9). *Cellular phones and cancer risk*. (National Cancer Institute) Retrieved July 4, 2019, from

Cancer causes and prevention: https://www.cancer.gov/about-cancer/causes-prevention/risk/radiation/cell-phones-fact-sheet

NIH. (2019, June 26). *Environmental Carcinogens and Cancer Risk.* Retrieved from National Cancer Institute: https://www.cancer.gov/about-cancer/causes-prevention/risk/substances/carcinogens

Oracle. (n.d.). *ava Telephony API (JTAPI).* (Oracle) Retrieved July 7, 2019, from Oracle: https://www.oracle.com/technetwork/java/jtapi-136088.html?printOnly=1

O-RAN Alliance. (n.d.). *Operator Defined Next Generation RAN Architecture and Interfaces.* (O-RAN Alliance) Retrieved July 6, 2019, from https://www.o-ran.org/

Penttinen, J. (2015). EMF - Radiation Safety and Health Aspects. In J. Penttinen, *The Telecommunications Handbook* (p. 956). Wiley.

Penttinen, J. (2015, November 20). *Mobile Generations Explained.* (Interference Technology) Retrieved July 2019, 2019, from https://interferencetechnology.com/mobile-generations-explained/

Penttinen, J. (2017, September 2). *5G testereihin monta kisaajaa (Finnish article on 5G testers with English summaries).* Retrieved from Uusiteknologia: https://issuu.com/uusiteknologia.fi/docs/2_2017/42

SDX Central. (2019). *How 5G SDN Will Bolster Networks.* Retrieved from Definitions: https://www.sdxcentral.com/5g/definitions/5g-sdn/

Sherman, J. (2013, July 26). *Spendy but indispensable: Breaking down the full $650 cost of the iPhone 5 .* Retrieved from Digital Trends: https://www.digitaltrends.com/mobile/iphone-cost-what-apple-is-paying/

Swisscom. (2019, March 27). *5G Mobile Technology Fact Check.* (Swisscom) Retrieved July 4, 2019, from Press Release: https://www.swisscom.ch/en/about/news/2019/03/27-5g-mobile-technology-fact-check.html

Tampere University of Technology. (n.d.). *Positioning and location-awareness in 5G networks.* (Tampere University of Technology) Retrieved July 3, 2019, from http://www.tut.fi/5G/positioning/

Wagner, M. (2019, February 12). *AT&T Building 5G Network on an Open Source Foundation*. (LightReading) Retrieved July 6, 2019, from https://www.lightreading.com/open-source/openstack/atandt-building-5g-network-on-an-open-source-foundation/d/d-id/749405

What is REST. (n.d.). Retrieved July 7, 2019, from REST API tutorial: https://restfulapi.net/

WHO. (2019, July 31). *Electromagnetic fields (EMF)* . Retrieved from EMF: www.who.int/peh-emf/en/

INDEX

Made in the USA
Middletown, DE
14 January 2020